P9-AOP-859

LINEAR PROGRAMMING

An Introductory Analysis

LINEAR

N. PAUL LOOMBA

Professor of Management
Lehigh University

McGRAW-HILL
BOOK COMPANY

New York
San Francisco
Toronto
London

PROGRAMMING

An Introductory Analysis

LINEAR PROGRAMMING

Copyright © 1964 by McGraw-Hill, Inc.   All Rights Reserved.
Printed in the United States of America.   This book, or
parts thereof, may not be reproduced in any form without
permission of the publishers.
Library of Congress Catalog Card Number 64-18902

38695

To

J. L. Loomba

Mary Adams Loomba

Caroline and Pierpont Adams

Preface

A new dimension has been added to the theory and practice of management during the last two decades. It is characterized by the availability and application of mathematical and quantitative models for purposes of analyzing and solving management problems. Most of the theoretical literature in, and practical applications of, this new dimension have come from pioneers in the field of operations research and/or management science. Having realized that the descriptive aspects of management had already reached a plateau of development, management scientists proceeded to evolve a theoretical framework which has given new life and vigor to this field. As a result, we now have access to a number of powerful optimizing methods and techniques which are helpful in solving a wide range of business and management problems. The area of inventory management, for example, has been subjected to intensive theoretical as well as practical research. Inventory models with different assumptions regarding demand and lead-time conditions have been built to solve the ever-present problems of "how much" and "when" to order or produce. Other representative examples include allocation models, replacement models, waiting-line models, and the like. Of these, allocation models are of special interest to the economist, for their central theme is the same as that of economic theory, namely, the allocation of limited resources to competing candidates or activities.

Linear programming is a subclass of allocation models. It is a method of allocating scarce resources to competing activities under the assumptions of linearity. In linear-programming problems, both the objective function and the constraints are assumed to be linear. In other words, linear programming deals with problems whose structure is made up of variables having linear relationships with each other. Within the framework of such allocation problems, linear programming is used either to maximize or to minimize a given objective function.

The credit for developing the first general method for solving linear-programming problems is usually given to Prof. George B. Dantzig. The first formal presentation of the simplex technique was in the form of a paper published by Professor Dantzig in 1951. Although a number of modifications and extensions of the simplex method have been developed since then, the original technique remains the most general approach to the solution of linear-programming problems. Its importance is further enhanced when we realize that specialized forms of this basic simplex method, such as the transportation and assignment models, lend themselves to some important industrial applications.

Linear programming is only one aspect of what has been called a "systems" approach to management wherein all programs are designed and evaluated in terms of their ultimate effects on the realization of business objectives. This approach recognizes the multiplicity of objectives in decision making and identifies the dangers of suboptimization. In addition to its emphasis on the overall objectives of the business organization, this approach is characterized by the fact that it views business problems in terms of building, manipulating, solving, testing, and applying decision models. Furthermore, it employs a quantitative analysis distilled from and based on the descriptive and qualitative aspects of management problems. That this approach is yielding significant dividends is evidenced by its increasing acceptance and use in industrial, government, and academic circles. A number of universities and colleges, for example, have already introduced courses in the field of operations research and management science.

Colleges of business administration, in particular, are developing their curricula with a three-way emphasis. First, a set of courses is designed to provide theoretical foundations based on economic theory, statistics, operations research, and behavioral sciences. Second, functional emphasis is provided through such courses as production, finance, and marketing. Third, there are courses which acquaint the students with such institutional factors in our society as legal and governmental aspects of management. The movement exemplified by this three-way emphasis is beginning to gain momentum. If the present trend continues, a course in operations research or linear programming will probably be an integral part of the business-administration curriculum in the very near future.

Linear programming may appear to be of limited value, owing to its reliance on the assumption of linearity. However, a number of practical problems are of such a nature that the assumption of linearity is close enough to reality. An understanding of linear programming can provide the executive with a powerful tool for decision making. Furthermore, in so far as linear programming is an integral part of management science, a knowledge of this field orients the student toward analytical thinking. In view of the competitive conditions prevailing in business and industry today, business-administration as well as engineering students must be exposed to the philosophy, techniques, and economic interpretation of linear-programming methods.   Although a number of books in the area of management and industrial engineering have dealt with this subject, most of these seem to get involved in mathematical exercises beyond the scope of many students.   The use of mathematics, as such, is perhaps desirable for giving the student a clear insight to linear programming. It should, however, be deferred until the time when a student can grasp the meaning, the mechanics, and the exact nature of basic linear-programming problems.   In the opinion and experience of this author, two things stand in the way of increased student understanding and acceptance of this area.   First, no text has attempted to integrate clearly the various approaches, such as the systematic trial-and-error method, the vector method, and the simplex method, to the solution of simple linear-programming problems.   As a result, the student is not able to integrate his visualization process with his computational type of learning.   Second, students seem to face a learning barrier because no presentation of the simplex method has made an explicit and persistent attempt to tie together the linear equations representing the problem, their physical relevance, and the economic interpretation and policy implications of the optimal solution.   This author has attempted to remove these impediments by emphasizing both the mechanics and the economic interpretations of linear programming throughout this text.

The purposes of this book, then, are to identify the place of linear-programming problems within the broad field of operations research, to provide a clear understanding of the simplex algorithm through the solution of linear-programming problems by various methods, and to explain the relationship between the simplex, transportation, and assignment models.   The text will have achieved its objective if it acts as a catalyst to a broader understanding, acceptance, and use of quantitative models in dealing with business and management problems.

This book has been written primarily for the beginning student in linear programming.   Valuable suggestions from many of my students have, indeed, become part and parcel of this volume.   In particular, I wish to thank Herbert E. Ehlers, James M. O'Brien, Jr., and Captain Frederick G. Tripp.

I am deeply indebted to Dr. Alan S. Foust, Dean of the College of Engineering, Lehigh University, for his valuable suggestions on the organization of this book.   Particular thanks are due to Dr. Elmer C.

Bratt, Head of the Department of Economics; Dr. Herbert M. Diamond, Acting Dean of the College of Business Administration; and Dr. Glenn J. Christensen, Provost and Vice President of Lehigh University, for providing the time and encouragement to complete this project.

Finally, I wish to acknowledge the help of Mrs. Charlotte Bradley, Mrs. Helen Green, and Mrs. Margaret Clauser, who patiently typed several revisions of the original manuscript.

N. Paul Loomba

Contents

Spatial

appendix

I

THE MEANING OF LINEARITY 241

appendix

II

A NOTE ON INEQUALITIES 250

appendix

III

A SYSTEM OF LINEAR EQUATIONS HAVING A UNIQUE SOLUTION 253

Linear Programming and Management

1.1 INTRODUCTION

Linear programming, although of recent origin, has already demonstrated its value as an aid to decision making in business, industry, and government. Determination of facility or machine scheduling, distribution of commodities, determination of optimum product mix, and allocation of labor and other resources are but a few examples of the type of problems which can be solved by linear programming.* Briefly, linear programming is a method of determining an optimum program of interdependent activities in view of available resources. The term *linear* implies that all relationships involved in the particular problem which can be solved by this method are linear.† The term *programming* refers to the process of determining a particular program or plan of action.

A linear-programming problem arises whenever two or more *candidates* or activities are competing for limited *resources* and when it can be assumed that all relationships within the problem are linear. For example, let us assume that we are to determine a production "program" involving three different products, say X, Y, and Z, each yielding a specific profit contribution per unit. Assume further that the manufacture of each product requires some given processing time in each of,

---

* Robert O. Ferguson and L. F. Sargent, "Linear Programming," pp. 10–11, McGraw-Hill Book Company, New York, 1958.
† See Appendix I for the meaning of *linear* relationships.

say, three different manufacturing departments.*   The products X, Y, and Z, in this case, are the *candidates*, and the three manufacturing departments are the *resources*.   Obviously, we want to design that program which will maximize contributions to profit.   This type of objective is usually quantified and becomes what is called an *objective function*. Assumption of linearity implies that it is a *linear* objective function.

In so far as it is assumed that the current capacity of the three manufacturing departments is not to be expanded during the given time period, the capacity of the resources, although known, is limited.   This type of limitation or specification is expressed by saying that a set of *structural constraints* is given in the problem.   Assumption of linearity implies that such a set is a set of *linear* constraints.   Furthermore, a production program, by definition, must be such that a particular *candidate* is either included in or excluded from the program.   This means that negative production, which has no physical counterpart, is not permitted in the solution of a linear-programming problem.   This obvious fact is made an integral part of the linear-programming problem by stating a set of so-called "nonnegativity constraints."

The above remarks regarding some of the terminology employed in describing a typical linear-programming problem are only for introductory purposes.   The exact nature of linear-programming problems and their solutions will become clear as we cover the material presented in Chapters 2 through 7.

The main purpose of this book is to present various methods of solving linear-programming problems and to provide economic interpretation, wherever possible, of the various components of the solution stages.   However, to provide the reader with a broad picture of the place occupied by linear programming in the area of operations research and/or management science, this chapter first presents a brief history of the evolution of modern management (Section 1.2).   An attempt is then made to define operations research and to identify its role in decision making (Section 1.3).   This is followed by a discussion of the process of decision making (Section 1.4) and a brief discussion of models and model

* *Process Time Data*

Department	Product			Capacity constraint per time period
	X	Y	Z	
Cutting..........................	10.7	5	2	2,705
Folding..........................	5.4	10	4	2,210
Packaging........................	0.7	1	2	445
Profit contribution per unit.........	$10	$15	$20	

building (Section 1.5). The last section of this chapter is devoted to a brief description of some well-known operations-research models. Some familiarity with these topics will enable the reader to understand the role and limitations of linear programming.

1.2   EVOLUTION OF MODERN MANAGEMENT

The Functional Approach

Management may be defined as the *process* of coordinating available resources to achieve economic, political, social, and other objectives. In this sense, the "practice" of management is as old as the history of man, for one cannot conceive of any practical set of circumstances wherein the need for employing available resources to accomplish certain objectives is not present. Nevertheless, for purposes of providing a concrete starting point for describing the evolution of management, it is usually claimed that it was during the period of the industrial revolution that the practice of scientific management first produced any significant results. As a matter of fact, the very development of the factory system of production may be ascribed to the application of scientific management based on the principles of division of labor and specialization. In retrospect, we can also observe a parallel development in the "theory" of management, exemplified by the writings of Charles Babbage in the early part of the nineteenth century and later followed by the work of men such as Frederic W. Taylor, Henri Fayol, Frank Gilbreth, and Henry Gantt.

Inevitably, with the passage of time, the developing theory and the increasing practice underwent a process of action and reaction. This gave birth to what is now known as the *scientific-management* movement. Operations research, management science, and associated theoretical developments in the field of management are the present-day culmination of this movement.

Taylor, Gilbreth, Gantt, and other pioneers of scientific management observed that, during most of the nineteenth century, considerable advances had been made in the design and utilization of specialized machines and tools. This phenomenon, furthermore, had been accompanied by an increased productivity on the one hand and the development of larger business organizations on the other. These pioneers soon recognized that, in order to gain further improvements in economic operations, the principle of specialization, already operating in the manufacturing activities, should be extended to the sphere of management activities. Taylor worked on this idea and developed what is known as

Taylor's functional organization. This particular form of organization was based on the argument that if the worker was to get expert advice in the performance of his varied tasks, such advice, of necessity, must come from several different specialists, for there are natural limitations on the capabilities of any single person in terms of becoming a specialist in more than one area. Taylor, therefore, suggested that production activities be organized in terms of functions to be performed, and that each worker be under the supervision of, and accountable to, more than one foreman. * Since this form of organization violated the principle of unity of command, it did not prove to be of lasting value in practice. However, it did provide the impetus for the development of the line-and-staff type of organization which is the predominant form of business organization today.

The purpose of the line-and-staff type of organization was, obviously, to provide the advantages of Taylor's functional emphasis without sacrificing the unity-of-command idea. Along with this new form of industrial organization, continuous efforts were made to evolve better methods for solving other management problems. Subsequent developments in the field of management show that it was the functional specialization of the separate parts, rather than the management system as a whole, that received foremost attention from management theoreticians and practitioners during the early decades of this century. George D. Babcock, for example, developed mathematical formulas in 1912 for determining economic lot sizes.† Merit rating systems and techniques of job analysis were developed during World War I. Frank and Lillian Gilbreth made notable contributions to the psychological aspects of management and developed principles of motion study. Henry Gantt introduced the use of schematic charts for purposes of scheduling and controlling production. To this day, most of the visual devices used to control production are based on the concepts contained in Gantt charts.

The increased use of industrial cost accounting, better methods of budgeting, standard costing, long-term forecasting, and work simplification are some of the other developments which took place during the 1920s. Statistical quality control, based on the theory of probability, was introduced during the 1930s. Walter Shewhart published his famous paper "The Economic Quality Control of Manufactured Products" in

---

* In the specific organization plan advanced by Taylor there were eight foremen exercising supervision over the workers. Of these, four foremen controlled the "planning" activities of time and cost, instruction, discipline, and order of work, while the other four supervised the "performance" activities of speed, progress, inspection, repair, and set-up.

† George D. Babcock, "Taylor System in Franklin Management," p. 125, Engineering Magazine Company, New York, 1917.

1930 and laid the foundation for the modern use of statistics in operations analysis. The introduction of work sampling for determining work standards was another important addition to the increasing number of management techniques being developed during the middle and late thirties. In summary, the industry by now had access to a number of rather refined tools of management. With the simultaneous use of these new techniques, business and industrial organizations were able to run their production systems more economically. A natural extension of this development was a further increase in the size of the business organization, and, as we shall discuss later, there arose a new class of problems—the so-called "executive-type" problems. These problems, then as now, are the result of the need for reconciling the often conflicting objectives of the different components of a large organization.

The purpose of this very brief sketch of the evolution of management up to the beginning of World War II is twofold. First, it points up the fact that most of the present-day management methods are of rather recent origin. Second, it emphasizes that up to the early 1940s management techniques were developed by employing a "functional" approach and basically had the objective of solving functional problems. In other words, the main emphasis was on the objectives of the separate parts of the business enterprise, rather than on the overall objectives of the whole system.

The Systems Approach

The important developments following the early 1940s, in comparison with previous developments, may be distinguished by one specific characteristic, namely, the evidence of increasing emphasis on the overall objectives of the business enterprise. It is this philosophy that is contained in the so-called "systems" approach, according to which the business enterprise must be analyzed as a system made up of interconnected functional components. Thus, implicit in this approach is the conviction that before implementing any functional solution, one must examine its ultimate effect on the system. This type of orientation eventually led to the growth of operations research.

Operations research and the advent of high-speed computers are perhaps two of the most important developments of the period following World War II. It may be mentioned that the parallel development of operations research and computers is not a coincidence. One could have only a limited value without the other. It is the availability of high-speed computers that has made the economic solution of large-scale operations-research problems a reality.   Computers have also facilitated the analysis and prediction of the behavior of complex production and economic systems by employing simulation models.*

These developments and their modifications and extensions, needless to say, will have further impact on the theory and practice of management in the years to come.

1.3   OPERATIONS RESEARCH

The availability of better management methods during the first four decades of this century brought profound changes to the business scene. Functional specialization, in both the technical and managerial fields, provided the means and opportunities for realizing further economies of scale in the production process.   The net result was a phenomenal growth in the size of the business enterprise.   Since business enterprises were organized into separate components (e.g., divisions, sectors, and departments) having their own specific goals, there was, and under similar circumstances always will be, the possibility of potential conflict among the goals of the individual components.   This state of affairs gave rise to the so-called "executive-type" problems which consider the effectiveness of the organization as a whole in view of the conflicting goals of its component units.   Let us illustrate by considering the problem of inventory control.

The question of inventory levels is a typical executive-type problem wherein different functional departments of a manufacturing enterprise have conflicting objectives.   The sales department, in general, would like to have a large inventory of different products to achieve uninterrupted and satisfactory customer service.   This clashes with the objective of the finance department, which would like to keep the inventory investment as low as possible.   Similarly, the production department may wish to manufacture in large lot sizes to minimize production costs, but this would give rise to increased in-process inventories and higher working-capital requirements.   The problem, therefore, is to develop an inventory policy which will minimize the total costs associated with the different requirements of the functional departments.   A number of similar

---

* Any time we attempt to approximate the behavior or characteristics of an object or subject of inquiry by employing some device, we are building a simulation model. In particular, an analog model, including the dimension of time, represents a simulation model (see Section 1.5).   The training of astronauts, business games, and Monte Carlo techniques are some examples of simulation models.   See C. West Churchman, R. L. Ackoff, and E. L. Arnoff, "Introduction to Operations Research," p. 174, John Wiley & Sons, Inc., New York, 1957.

examples can be given to emphasize the existence of multiple and often conflicting objectives for both organizations and individuals.

Business demands on the discipline of management in the early forties were, therefore, different from those associated with purely functional problems. New methods of handling executive-type problems had to be developed. In addition, during World War II the need arose for solving some tactical and strategical military problems. These problems, in many respects similar to the newer problems of business management, were so complex that their solution required the pooled efforts of specialists from several disciplines. As a result, a body of knowledge consisting of certain management methods, mathematical tools and techniques, and decision models was developed. Linear programming is only one part of this body of knowledge, which has come to be known as operations research and/or management science.

What is operations research? There is no unique answer to this question. One can find a number of definitions or explanations of operations research in the management literature. The subject of operations research, for example, is sometimes introduced by listing such topics as linear programming, queueing theory, Monte Carlo, and inventory models. These techniques were developed as a result of the operations-research effort and have come to be associated with the name operations research. There are, of course, other ways to define or explain operations research. A descriptive definition given by Morse and Kimball* is: "Operations research is a scientific method of providing executive departments with a quantitative basis for decisions under their control." Saaty† gives his favorite definition: "Operations research is the art of giving bad answers to problems to which otherwise worse answers are given." An attempt to explain operations research is sometimes made by listing such characteristics of its techniques as: (1) operations research employs a quantitative rather than qualitative approach and (2) its main mode of analysis is the building and using of symbolic models. Professor Dorfman claims that operations research is not a subject-matter field but an approach. To quote Professor Dorfman:‡

According to a survey conducted by the American Management Association, more than 40 per cent of operations analysts are engineers by training, another 45 per cent are mathematicians, statisticians, or natural scientists. It is only natural that the point of view in which these men are schooled should permeate operations

* Philip M. Morse and G. E. Kimball, "Methods of Operations Research," p. 1. John Wiley & Sons, Inc., New York, 1951.
† T. L. Saaty, "Mathematical Methods of Operations Research," p. 3, McGraw-Hill Book Company, New York, 1959.
‡ Robert Dorfman, Operations Research, *American Economic Review*, vol. 50, no. 4, p. 577, September, 1960.

research.   The essence of this point of view is that a phenomenon is understood when and only when it has been expressed as a formal, really mechanistic, quantitative model, and that, furthermore, all phenomena within the purview of science (which is probably all the phenomena there are) can be so expressed with sufficient persistence and ingenuity.   A second characteristic of men of science, amounting to a corollary of the first, is their preference of symbolic, as opposed to verbal, modes of expression and reasoning.   These characteristics I take to be the style of operations research, and I define operations research to be all research in this spirit intended to help solve practical, immediate problems in the fields of business, governmental or military administration or the like.

A working definition of operations research is advanced by Miller and Starr:* ". . . Operations research is *applied decision theory.* [It] uses any scientific, mathematical, or logical means to attempt to cope with the problems that confront the executive when he tries to achieve a thoroughgoing rationality in dealing with his decision problems."

Of the various definitions examined above, the applied decision-theory viewpoint is most general and broadest in scope.   Furthermore, considered in this fashion, operations research need not be limited to quantitative models.   Instead, it becomes both a guide and a tool in the process of decision making discussed in the next section.

1.4   DECISION MAKING

Making decisions is an integral and continuous aspect of human life. For child or adult, man or woman, government official or business executive, worker or supervisor, participation in the process of decision making is a common feature of everyday life.   What does this process of decision making involve?   What is a decision?   How can we analyze and systematize the solving of certain types of decision problems?   The answers to these and similar questions form the subject matter of *decision theory.* Decision theory is a rich, complex, growing, and interdisciplinary subject. Systematized procedures which, for example, facilitate allocation decisions comprise just one of the many fruitful products of research in this area.   Application of the Monte Carlo technique, waiting-line models, and replacement models are some other examples of the subject matter.

Some Views on Decisions and Decision Making

The universality and antiquity of decision making are obvious.   Many different disciplines have sought to explain the nature of decisions and the process of making decisions.   Philosophers, for example, have sought

* David W. Miller and Martin K. Starr, "Executive Decisions and Operations Research," pp. 103–104, Prentice-Hall, Inc., Englewood Cliffs, N.J., 1960.

to explain how a "good" decision is made. What is good, according to some philosophers, is determined by the value system of an individual. Given the tangible translation of an individual's value system, the individual proceeds to make a rational choice to achieve his goals. Economists have attempted to explain the consumer's decision to buy and the producer's decision to sell goods and services in terms of utility. Although the measurement of utility poses certain complex problems, the decision maker, according to economists, is supposed to proceed in a manner which will maximize his utility.

Psychologists and sociologists have seriously questioned the economist's concept of utility maximization as a criterion in making decisions. The argument is that decisions are not governed solely by market-determined utility. Habit, tradition, and other motivational causes also have considerable influence on the decision maker. Thus, one can find in the literature a number of interesting views on the nature of decisions and decision making.

Sequential Decisions

Other aspects of the nature of decisions and decision process are also worth exploring. For example, although making a decision implies coming to a conclusion or the termination of a process, the results of a particular decision may provide data for a subsequent decision. Thus decisions may be viewed as a means-end chain.* Vance makes the same point through what he calls the "Hegelian character" of decision making.† Thus, most decisions are actually parts of a sequence of decisions. This is especially true of managerial decisions in which the implementation of decisions and subsequent control are successful only if an effective element of feedback is ensured.

Because of the sequential nature of decision making, a decision should be made only after the effect of possible subsequent courses of action is considered.‡

Conscious Decisions

Decisions can be viewed as conscious and/or unconscious responses to a combination of external and internal stimuli. The distinction between

---

* Henry H. Albers, "Organized Executive Action: Decision-making, Communication, and Leadership," p. 203, John Wiley & Sons, Inc., New York, 1961.

† Stanley Vance, "Industrial Administration," p. 171, McGraw-Hill Book Company, New York, 1959.

‡ Burton V. Dean, Applications of Operations Research to Managerial Decision-making, *Administrative Science Quarterly*, vol. 3, no. 3, p. 415, December, 1958.

the unconscious and conscious responses is essentially in the dimension of time.   In the case of an unconscious response, such as recoiling from a hot kettle, the decision process takes an infinitesimally small amount of time.   What takes place in the human brain while such decisions are made is an interesting subject for the physiologists and psychologists. Some research on this aspect of decision making, which is essentially a part of the overall attempt to explain the human brain mechanism making decisions, has already been conducted.*   More important, however, are decisions which are the results of conscious responses.   For here the process of decision making takes place over a definite period of time, and thus analysis of the types of decisions made and an enumeration of the steps involved can be useful in developing a science of decision making. It is to assist the conscious type of decisions that most of the modern tools of quantitative analysis have been developed.

Managerial Decisions

Another way to classify decisions is to distinguish between managerial and nonmanagerial decisions.   In a typical business organization, for example, certain decisions at almost all levels of hierarchy are designed to affect the behavior of subordinates.   These are what may be called *managerial* decisions.   The decision of a worker to speed up his machine for a certain time, on the other hand, usually affects his own behavior only.   This type of decision, then, is an example of a *nonmanagerial* decision.

The Making and Implementation of a Decision

The implementation of a decision is another important consideration. If a decision is not put into effect in a manner that will produce the desired results, the very effort, time, and money expended in making the decision are wasted.   Thus the elements of acceptability and flexibility are extremely important in the two parts of the decision cycle: (1) the making of a decision and (2) the implementation of the decision.†

Analytical Decision Making

The analytical approach to decision making classifies decisions according to the amount and nature of the available information which is to be fed

* Albers, *op. cit.*, p. 221.
† Vance, *op. cit.*, p. 168.

as input data for a particular decision problem.   Since future implementation is an integral part of decision making, available information is classified according to the degree of certainty or uncertainty expected in a particular future situation.   With this criterion in mind, three types of decisions can be identified.   First, there are decisions made when "future" can be predicted with *certainty*.   In this case, the decision maker *assumes* that there is only one possible future in conjunction with a particular course of action.   This type of situation may be illustrated by considering a production-scheduling problem.   Suppose that in a machine shop three jobs are to be completed by utilizing three different machines, and that the time required to complete each job on each machine is known. The problem is to assign these jobs to machines with the objective of minimizing production time.   All the decision maker has to do in such a case is to consider the various combinations in which jobs can be assigned and to calculate the production time associated with each combination. Of these, the decision maker chooses the one with the minimum production time.

Second, there is decision making under conditions of *risk*.   In this case, the future can bring more than one state of affairs in conjunction with a specific course of action.   Consider, for example, that we decide to place a bet on the outcome of one flip of a coin.   Obviously, there are only two possible states of the future—the coin will show either head or tail. Furthermore, in this case, it is possible to make an a priori statement about the probability of obtaining each future state.   Similarly, one can state the probability of drawing an ace from a deck of cards, throwing 1, 2, etc., on the roll of a die, or drawing a green chip from a bowl which contains, say, two green and three red chips, all on an a priori basis.   In other cases, assuming conditions of stability and based on *empirical* results, probability statements can be made in connection with the outcomes of some business events.   For example, marketing research may indicate the respective probabilities of attaining different levels of future sales.   In any case, the characteristic of a future involving risk is that the probability with which a particular state of affairs will occur can be stated—either on an a priori basis or on the strength of empirical data.

Third, there is decision making under *uncertainty*.   In this case, a particular course of action may face different possible futures, but the probability of each occurrence cannot be estimated objectively.   In the final round of a table-tennis tournament, for example, we may sincerely refuse to project the chances of any one player winning the match. Here there are two possible events for each player, win and lose, but because of insufficient supporting data, we may not be able to attach any probability values to the outcome.   This simple illustration provides an example of an uncertain future.

Of the three types of futures discussed above, linear-programming problems fall in the category for which the future can be predicted with *certainty.*

Regardless of the type of future faced in a given situation, the analytical approach to decision making follows the logical sequence of steps which will now be explained.

1. *Formulation and statement of objectives.* In a typical managerial problem, the manager is required to make decisions regarding the quality and quantity of input (in the form of human, physical, and financial resources) in order to achieve a certain amount of output. The achievement of this output, the manager believes, is essential for bringing about a certain state of affairs which will enable him to realize his goals or objectives. The first step, therefore, is the formulation and precise statement of objective(s) which may or may not be amenable to quantification. If a natural scale for measurement is available, the requirement to quantify is easily satisfied; otherwise the decision maker may have to utilize such tools as the standard-gamble* technique.

In reference to a typical linear-programming problem, an objective which is to be either maximized or minimized is specified. For example, if the manager assumes that the profit contributions per unit of two different products (utilizing the same but limited resources) are fixed for the planning period, his objective may be to maximize the total profit contribution which will result from some product mix of these products (see Table 2.2).

2. *Identification and particularization of pertinent variables.* This step is the core of what is called "model building." Having identified the objective variables, the decision maker makes a choice of those variables whose future magnitudes will affect the degree of achievement of his objective(s). The degree of such influence(s) is then "particularized" by some quantitative relationship, which is usually a mathematical model. If the problem is such that mathematical formulation and manipulation are difficult, expensive, and time-consuming, the decision maker may have to utilize such tools as simulation.

In reference to a typical linear-programming problem, the variables affecting the specified objective are explicitly stated. Furthermore, the degree of their influence on the stated objective can easily be particularized.†

---

* John von Neumann and O. Morgenstern, "Theory of Games and Economic Behavior," Princeton University Press, Princeton, N.J., 1947.

† For the linear-programming problem given in Table 2.2, a quantitative statement of the objective function is obviously: maximize $10X + 15Y$, where X represents some unknown quantity of product A, and Y represents some unknown quantity of product B.

3. *Formulation of available strategies.* In a number of cases, the result of step 2 brings the decision maker to the point at which the use of mathematics will identify immediately the optimum strategy in a given problem. There are situations, however, in which a more time-consuming process of formulating a number of specific strategies may have to be undertaken. In such cases, the decision maker simply lists the various combinations of resource factors under his control, each such combination representing a specific strategy.

A typical linear-programming problem, as we shall discover later, is characterized by the fact that an infinite number of strategies are available. In the linear-programming problem of Table 2.2, any combination of products *A* and *B* is an available strategy. Needless to say, the decision maker simply cannot list all the available strategies. Instead, he utilizes various linear-programming methods which lead him, step by step, to successively better strategies, until an optimum strategy is identified.

4. *Prediction of the payoff.* In addition to listing the possible combinations of resource factors under his control, the decision maker may, in a number of situations, be required to predict the possible strategies of his competitors and the possible future magnitudes of some pertinent variables which are not controlled by him. Here, again, mathematical and statistical models are employed. The decision maker must now predict the *payoff*, which is his estimate of the degree of achievement of his objective(s) under various possible combinations of controllable and noncontrollable variables.

In linear-programming problems, all variables are assumed to be controllable. Furthermore, the competitor's strategies are assumed to have no effect on the payoff, which can be predicted with *certainty* for each possible strategy. For example, in the linear-programming problem given in Table 2.2, the payoff per unit is simply the profit contribution per unit associated with each product.

5. *Making the decision.* Next, the decision maker analyzes the information obtained as a result of implementing the earlier steps and makes a "rational" choice of a specific strategy. In other words, the choice is made according to a particular decision criterion. Other things being equal, this choice will yield a maximum degree of achievement.

In linear-programming problems, the analyst simply applies a test of optimality for identifying the optimum solution. The optimum solution represents that specific strategy which either maximizes or minimizes the objective function.

Thus, regardless of such specific characteristics of the decision problem as "assumed certainty" or "probabilistic" future, linearity or nonlinearity, and "static" or "dynamic" considerations, the analytical approach described above can be uniformly followed. The point to emphasize here is that such tools as linear programming have built-in mechanisms which guide and carry out the various steps of analytical decision making. However, the analyst must feed the pertinent data in the required form.

1.5   MODELS AND MODEL BUILDING

The logical analysis of all decision problems is based on the concept of models and model building. To build a model, we have to use some device to represent an object or subject of inquiry. This device may be schematic, physical, symbolic, or a combination of these. The flow-process chart showing the complete processing of a product, for example, is a *schematic* model to represent different operations and such other activities as storage, inspection, transportation, and delay. A toy airplane or a three-dimensional small-scale model of a new building is a *physical* model. An organization chart is a *graphic* (block-type) model showing authority-responsibility relationships in an organization. The equation $Y = a + bX$, representing a straight line, is a *symbolic* or *mathematical* model. In short, a model embodies our attempt to represent reality. The model is built with the purpose of enhancing our understanding of reality.

The reality, however, is so complex that often we cannot visualize and understand it completely. Furthermore, even if we do understand it, our attempts at representation may succeed only partially, because the tools and means needed to represent it may not be developed or available. For example, dynamic and transient models of the economic firm are seldom constructed, because the discipline of mathematics, as yet, is not equipped to handle such models. Thus, by definition, there are some inherent weaknesses in the use of models to represent reality. This, however, does not mean that we should reject the approach of model building. On the contrary, analytical model building is not only sound but the best available approach to a study of decision problems. It is much easier, less costly, and less time-consuming to obtain information from models than from experimentation with the reality that the model represents. Furthermore, the main advantage of a "tested and proved" model is that once a general model is set up, it can be used to solve a whole class of similar problems. All the analyst has to do is recognize the nature of the given problem and see if it may be represented by a particular proved model. Thus, as we shall see later, all maximization and minimization problems having the characteristics of a strictly defined linear-programming problem can be solved by the general model of the simplex method.

The purpose of a model is to describe, explain, or predict the performance of a system. To be most useful, the model should explain and predict the behavior of individual components of a system, the response of the system to changes in one or more of its components, and the effect on the system of internal as well as external disturbances. After the model has been constructed, it should be tested for applicability. A successful test of applicability transforms the model into theory. However, a model need not be 100 percent applicable. The real issue is the degree of usefulness of the model. Thus, the assumption of linearity in linear-programming models may not be completely applicable to real business problems, but they are nevertheless extremely useful in solving certain types of problems.

Types of Models

The extensive application of models in the field of operations research makes it desirable that the reader be familiar with different classifications of models. One classification of models is based on the degree of abstraction* (see Figure 1.1).

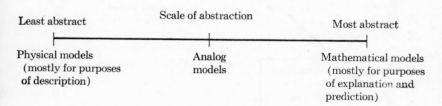

Figure 1.1   Classification of models according to degree of abstraction.

Physical models are those models which have the appearance of the real thing. Examples of such models are children's toys, sculptures, photographs, and paintings. Physical models are easy to observe, build, and describe, but they are difficult to manipulate and not very useful for purposes of prediction. Depending upon their specific purposes, physical

---

* C. West Churchman, R. L. Ackoff, and E. L. Arnoff, "Introduction to Operations Research," chap. VII, John Wiley & Sons, Inc., New York, 1957.

models are scaled up or down.   The use of three-dimensional models and templates in plant layout is a well-known application of physical models.

Analog models are more abstract than physical models, as they do not resemble the real object.   They are built by utilizing a set of properties or

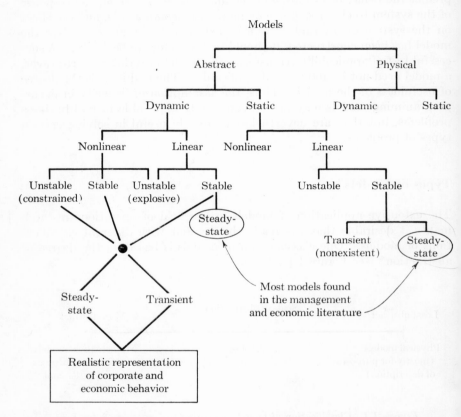

Figure 1.2   Classification of models.   (*From J. W. Forrester, "Industrial Dynamics," p. 49, The M.I.T. Press, Cambridge, Mass., and John Wiley & Sons, Inc., New York, 1961.*)

scheme different from that which the object of inquiry possesses.   For example, different colors on a map correspond to different character-istics—blue for water, brown for deserts, etc.   Graphs of time series, sales volume, stock-market changes, and the like are other examples of analog models.   Analog models are easier to manipulate and more general than physical models.

Mathematical models or symbolic models are most abstract. They employ a set of symbols to represent the system of inquiry. They are general rather than specific. Further, they are precise and may be manipulated exactly by utilizing the laws of mathematics. A straight line, for example, is represented by an equation $Y = a + bX$. And this equation represents *any* straight line with a given intercept and slope.

Professor Jay W. Forrester gives a classification of models which is very useful in recognizing the role and place of linear-programming models (see Figure 1.2).

Linear-programming models, in so far as they assume a *certain* future and are mathematical models, fall under the category of abstract, stable, static, and linear models.

1.6   SOME OPERATIONS-RESEARCH MODELS

As progress in the field of operations research continues, analysts are formulating operations-research models to solve certain *classes* of problems. Of the models developed so far, the following will be briefly described:

1. Allocation models
2. Inventory models
3. Waiting-time models
4. Replacement models
5. Maintenance models

Allocation Models

Allocation models are used to solve that class of problems in which a number of candidates or activities are competing for limited resources. The resources are limited in two ways. First, in a given time period, there may be a limit on the quantity, beyond which resources cannot be purchased or employed. Second, they may be limited in the sense that within the boundaries of the problem each candidate usually cannot be allocated the given resources in the most efficient way on an individual basis. For example, let us assume that during a particular time period a firm is planning to manufacture three different products X, Y, and Z, each yielding a specific contribution to overhead and profit. Assume, further, that two different machine processes (resources) are required in manufacturing each of these products. The allocation problem is to determine a program of production or a product mix which will maximize not the individual contribution of a given product but the overall effectiveness of the production program. Mathematical programming, linear and nonlinear, is one of the methods used in solving such problems. Other variations of allocation problems arise in determining the choice of candidates to utilize given resources or in finding suitable resources to satisfy the demands of a given set of candidates.

Inventory Models

Inventory models deal with that general class of problems in which something is stored to meet future demand. Essentially, inventory models answer the questions "How much?" and "When?" in relation to the quantity and time of procurement and/or production for items of interest. Associated with any inventory problem are two sets of costs. As the level of average inventory over a particular time period increases, one set of costs increases while the other decreases. For example, for a known demand rate and a known lead time, and with no discounts for quantity purchases, ordering costs decrease while storage costs increase as a function of average inventory levels. The determination of the economic order quantity requires the balancing of these two types of costs.

For purposes of analysis and eventual development of an inventory model, therefore, one must identify all such costs, determine their relationship to various inventory levels, and measure them in relation to different strategies. Once the model has been developed, it can be solved by employing analytical, numerical, or simulation methods. This solution, in effect, yields the particular strategy which will result in an optimal value of the selected measure of overall effectiveness.

Waiting-time Models

Waiting-time models are a special case of inventory models. Here, the problem is such that a number of facilities (men or machines) are provided to serve a number of arrivals (customers or products requiring service). If a perfect balance does not exist between the service facilities and the customers, waiting is required of either the service facilities or the customers. As in the inventory problem, depending on the number of service facilities in relation to the pattern of service requirements, two opposing sets of costs are associated with a given level of waiting time. The decision problem is to provide a specific number of service facilities and to control the arrival rate of the customers in order to minimize the sum of these two sets of costs. Operations-research models have been developed to solve this type of problem.

Replacement Models

Customarily, replacement problems are considered in two categories. The first deals with equipment that deteriorates with time. Equipment may lose efficiency as a result of either use or the appearance of better equipment on the market. Lathes, drilling machines, planers, and electronic devices are representative of the type of equipment that deteriorates owing to use or obsolescence. The second category deals with items that have a more or less constant efficiency with time. When they fail, they do so suddenly and completely. A typical example of a problem in the second category is the decision as to a replacement policy for light bulbs.*

Methods of analysis for formulating replacement policies with regard to these two categories are dissimilar because of the different nature and cost behavior of the equipment involved in each category. For equipment that deteriorates with time, for example, the analyst must consider, among other costs, the operating and maintenance costs that increase with time. It is necessary, when considering problems in this category, to decide upon an *optimum* interval of time after which the present equipment should be replaced with another candidate.

The category dealing with items that fail suddenly and completely calls for an analysis in which some sort of mortality distribution is predicted for the items in question. Based on this distribution, the replacement policy has the objective of minimizing costs by determining a certain interval after which all items are replaced (within this interval individual items that fail are replaced immediately). Although this category of problems deals with decreasing, constant, and increasing probabilities of failure with time, a commonly used illustration is that of group replacement of light bulbs having an increasing probability of failure. In particular, the problem is to determine some interval of time such that the combined cost of replacing individual items within this interval and replacing *all* items at the end of the interval is minimized.

Maintenance Models

Maintenance problems can be regarded as a special case of replacement or inventory problems. They are replacement problems in the sense that

* Maurice Sasieni, A. Yaspan, and L. Friedman, "Operations Research: Methods and Problems," p. 108, John Wiley & Sons, Inc., New York, 1960.

maintenance usually involves replacing parts after they fail to function effectively. They are inventory problems to the extent that maintenance crews or facilities are stored to serve the maintenance needs of the future. Maintenance problems, like inventory problems, involve two opposing sets of costs. One results from machine or facility breakdown; the magnitude of the costs in this set increases as the *average idle machine time* increases. The other set of costs results from measures adopted to decrease the average idle machine time; these costs increase as the average idle machine time decreases. Thus, the two sets of costs move in opposite directions as different levels of the average idle machine time of a particular machine are considered. The problem is to design and operate a maintenance program so that the sum of these two sets of costs is minimal.

1.7   SUMMARY

Management theory, during the last two decades, has been shifting its emphasis from a purely functional approach to a systems approach toward solving management problems. As a result, a number of so-called operations-research models have been developed to assist in the solution of management problems. Linear programming comprises one such model.

Of the different types of models discussed in this chapter, we can say that linear programming falls in the category of static, stable, and linear models. Linear programming is essentially a method of determining an optimum program or "mix" of the candidates or interdependent activities which are competing for limited resources, under assumptions of linearity.

A linear-programming problem consists of three parts. First, there is a linear objective function which is to be either maximized or minimized. Second, there is a set of linear constraints which contains the technical specifications of the problem in relation to the given resources or requirements.* Third, there is a set of nonnegativity constraints—since negative production has no physical counterpart.

The method of linear programming is such that it has a built-in mechanism for carrying out the essential steps involved in solving problems having the characteristics just stated. The next chapter is devoted to the graphical method of solving linear-programming problems.

---

* Since these constraints specify the structure of a given linear-programming problem, they are usually referred to as *structural constraints*.

The Graphical Method

2.1 INTRODUCTION

Linear programming, as stated in the previous chapter, is a method of solving the type of problem in which two or more "candidates" (activities) compete for limited resources. The allocation of resources is determined with the objective of either maximizing or minimizing a linear objective function (say profit or cost, respectively). How is this allocation determined? What are the different methods for handling such problems? What are the actual mechanics of these methods? We shall answer these questions by considering a typical linear-programming problem and solving it by (1) the graphical method, (2) the systematic trial-and-error method, (3) the vector method, and (4) the simplex method. The purpose of concentrating on the *same* problem is to enable the reader to grasp the relationships among the different solution stages involved in the various methods for solving linear-programming problems.

Of the above-mentioned approaches to linear programming, the simplex method represents the most general and powerful. Before presenting the simplex method, which will be covered in Chapters 6 and 7, we shall discuss the graphical method (Chapter 2), the systematic trial-and-error method (Chapter 3), and the vector method (Chapter 5). This scheme is adopted for two reasons. First, linear-programming problems involving three or less competing candidates are more easily solved by these methods. Second, some familiarity with these methods is essential for understanding the mechanics and rationale of the simplex method.

2.2   THE PROBLEM

By means of some sharp bargaining with the union and the subsequent reduction of union "make-work" restrictions in his former contract, a small paper-towel manufacturer has created some spare capacity in each of his three main production departments: cutting, folding, and packaging.   For purposes of identification, three different sizes of paper towels currently produced by the company may be called products *A*, *B*, and *C*. Owing to its small size, the company can sell in the market all that it can produce at a constant price.   Management is inclined to be conservative and does not wish to expand production facilities at this time, although they do wish to utilize fully the present spare capacity.

The paper toweling is received from another manufacturer in large rolls.   These rolls are subsequently cut, folded, and packaged in the three sizes.   The pertinent manufacturing and profit information for each size of paper towel is summarized in Table 2.1.   The problem is to determine the most profitable "mix" of the products as an addition to present monthly output.

Table 2.1   *Process Time by Size and Department**

Department	Size			Constraint per time period†
	A	*B*	*C*	
Cutting..........................	10.7	5.0	2.0	2,705
Folding..........................	5.4	10.0	4.0	2,210
Packaging.......................	0.7	1.0	2.0	445
Profit contribution per unit..........	$10	$15	$20	

* Prepared by Mr. Rene W. Sopher, a management major (June, 1963) at the State University of Iowa, Iowa City.
† For example, the capacity constraint for the cutting department may mean that only 2,705 minutes per week of machine time is available in the cutting department.

The data contained in Table 2.1 represent the technical specifications of the three products.   The production of 1 unit of size *A*, for example, requires 10.7 units of processing time (say minutes) in the cutting department, 5.4 minutes in the folding department, and 0.7 minute in the packaging department.   When sold, product *A* yields a profit contribution of $10.00 per unit.   The total available capacity in the cutting department is 2,705 minutes.   Similar information is available in Table

2.1 for producing a unit each of products *B* and *C*.   Note that all these products *must* be processed through all three departments.

A quick examination of the data in Table 2.1 also reveals that product *A* is more "efficient" than product *B* in the use of folding capacity, whereas product *B* is more efficient in its demand on cutting capacity. Similar observations can be made about product *C*.   Different degrees of efficiency of products competing for the use of various available resources are typical in linear-programming problems.

2.3   THE GRAPHICAL METHOD (A TWO-DIMENSIONAL CASE)

A linear-programming problem in which either two or three candidates are competing for available resources can always be solved by the graphical method.   The reason is quite obvious.   Whereas we can easily construct and visualize a two-dimensional or three-dimensional space, our capability to graph and visualize a space containing more than three dimensions is limited.   For purposes of illustration, let us assume that our manufacturing company has arbitrarily decided to produce only products *A* and *B*.   The pertinent data for this problem are given in Table 2.2.   Our objective is to determine that mix of products *A* and *B* which will yield the maximum profit contribution.

Table 2.2   *Process Time by Size and Department*

Department	Size		Constraint per time period
	A	*B*	
Cutting......................	10.7	5.0	2,705
Folding......................	5.4	10.0	2,210
Packaging....................	0.7	1.0	445
Profit contribution per unit......	$10	$15	

The graphical method requires the translation of the above data into inequalities.*   Before we illustrate the graphical method, it is desirable that the reader gain familiarity with these inequalities and their physical interpretation.   For each "resource" in Table 2.1, a separate inequality

* See Appendix II.

can be written. If we let variables X and Y denote, respectively, the units of products A and B to be produced, the inequality for the cutting department is

$$10.7X + 5.0Y \leq 2,705$$

This inequality is simply an algebraic way of expressing the information given in Table 2.2 in connection with the cutting department. It gives algebraic expression to one of the three structural constraints in our problem. A descriptive translation of this inequality is: Each unit of product A requires 10.7 units of the cutting capacity, and each unit of product B requires 5 units of the cutting capacity; the maximum amount of A (that is, some specific value of X) and the maximum amount of B (that is, some specific value of Y) to be produced should be such that the total demand on the capacity of the cutting department does not exceed 2,705 units. Thus, any specific combination of values of X and Y that does not violate this and the other given constraints is a possible solution.

Let us assume that we wish to produce just 2 units of product A and 2 units of product B. This program ($X = 2$, $Y = 2$) can give us two types of information in connection with *each* resource: (1) the total capacity used by this program and (2) whether or not the capacity constraint has been violated. In this case, for example, the total cutting capacity used is $10.7(2) + 5(2) = 31.4$ units, and the remaining cutting capacity therefore is $2,705 - 31.4 = 2,673.6$ units. The constraint on the capacity of the cutting department obviously has not been violated. For each program, a similar check must be made of *all* resources.

As can easily be ascertained, $X = 2$ and $Y = 10$ is also a possible solution for this problem. Furthermore, this latter program results in a higher level of profit contribution than our first program ($X = 2$, $Y = 2$). However, there may be other programs which will yield larger profit contributions than the second program ($X = 2$, $Y = 10$). If so, we must discover them. This suggests that the search for a better program must continue until an optimum solution is determined. The optimum solution, in this case, will be a specific program which will yield the highest level of profit contribution without violating any of the structural constraints.

Transforming the Problem Data into Inequalities

Let variables X and Y denote, respectively, the units of products A and B to be produced. Then, the technical specifications of Table 2.2 can be transformed into the following inequalities:

For cutting capacity:

profit fns
16 x +15 y

$$10.7X + 5.0Y \leq 2{,}705 \tag{1}$$

For folding capacity:

$$5.4X + 10.0Y \leq 2{,}210 \tag{2}$$

For packaging capacity:

$$0.7X + 1.0Y \leq 445 \tag{3}$$

The objective in this problem is to determine at least one pair of values for X and Y (which satisfy all the structural constraints given by the above inequalities) resulting in the maximum possible level of the function $10X + 15Y$.

Graphing the Inequalities

The set of inequalities given above can easily be graphed. Inequality (1), for example, is graphed in Figure 2.1.

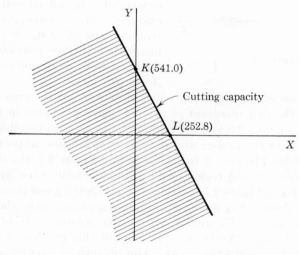

Figure 2.1

To obtain the X and Y intercepts for inequality (1), we proceed as follows:

Let $X = 0$; then

$$Y = \frac{2,705}{5} = 541 \qquad \text{point } K$$

Let $Y = 0$; then

$$X = \frac{2,705}{10.7} = 252.8 \qquad \text{point } L$$

Joining points K and L gives us a line whose equation is

$$10.7X + 5.0Y = 2,705$$

However, since we do not wish to plot the above equation, but rather the inequality $10.7X + 5.0Y \leq 2,705$, the region of interest is represented by the shaded area in Figure 2.1. Clearly, the shaded area also includes negative values of X and Y which mean negative production and have no real physical counterpart. In order to exclude any possibility of negative production, as mentioned in Chapter 1, a set of nonnegativity constraints is introduced. In our example, the nonnegativity constraints are $X \geq 0$, $Y \geq 0$; these mean that we are restricted to producing zero or more units of products A and B.

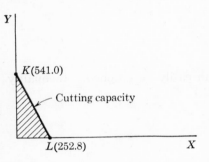

Figure 2.2

The addition of nonnegativity constraints restricts the area of possible solutions to the first quadrant of the XY plane as shown in Figure 2.2.

Similarly, Figures 2.3 and 2.4 represent the areas of possible nonnegative solutions for the folding and packaging capacities, respectively.

If we combine Figures 2.2 to 2.4, we obtain Figure 2.5, the shaded area of which represents the region of all possible solutions for our problem.

The drawing of Figure 2.5 is the first step in solving our problem by the graphical method. Any point in the shaded area and/or on the boundary of the shaded area of Figure 2.5 is a possible solution. In other words, there are an infinite number of solutions for this problem if we assume divisibility of the production units. Our objective is to pick at least one point (X, Y) from the shaded (feasible) area of Figure 2.5 which maximizes the profit function.

How can we proceed to accomplish this? It is at this point that we are guided by the profit function. If, somehow, we can graph the profit function on Figure 2.5, determine its direction of maximum increase, and start and keep on moving it in this direction, it will eventually touch some farthest point on the boundary of the shaded area.   This point, then, will give us a unique (or one of many possible) optimum and

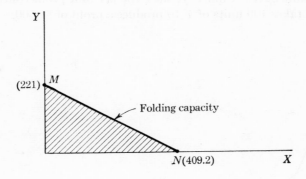

Figure 2.3

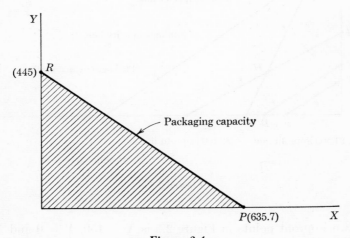

Figure 2.4

feasible solution.   The graphical method, as we shall illustrate below, accomplishes these things for us in a systematic fashion.

Isoprofit Lines

The graphing of the profit function may at first appear difficult, for the profit function $10X + 15Y$ is not in the form of an equation and hence

cannot be graphed.   But we can overcome this difficulty by asking our-
selves a simple question: How many units of product X *alone* (or product Y *alone*) are required to produce a profit contribution of, say, $1,500?*
Since profit contribution per unit of product X is $10, the answer is obvi-
ously 150 units of X.   Similarly, since the per-unit profit contribution of
Y is $15, it takes 100 units of Y to produce a profit of $1,500.   Thus, one

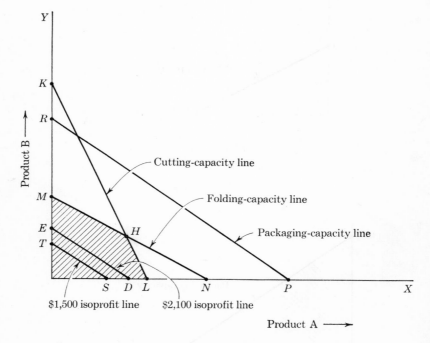

Figure 2.5

set of two isoprofit points in Figure 2.5 is $X = 150$, $Y = 0$ and $X = 0$,
$Y = 100$.   If we join these two points (S and T), we obtain the $1,500
isoprofit line (see Figure 2.5).   All points on this line, since they are
within the shaded area, represent feasible solutions, each giving a total
profit contribution of $1,500.   Thus, an isoprofit line is the locus of all
points (i.e., all possible combinations of X and Y) which yield the same
profit.

---

* Any arbitrary profit figure will work.   However, it is always simplest to pick a
profit number which gives an integer as an answer and which places the isoprofit line
within the shaded area.

In a similar fashion we could have drawn other isoprofit lines yielding different levels of profit contribution.   For example, line *DE* in Figure 2.5 represents the \$2,100 isoprofit line.   A comparison of lines *DE* and *ST* shows that the \$2,100 isoprofit line (*DE*) is parallel to the \$1,500 isoprofit line (*ST*) and is located farther away from the origin.   This was to be expected, since the per-unit contributions of the two products are fixed, and larger profit contributions result as we move away from the origin. A little reflection will show that we should keep on drawing such isoprofit lines for higher profits so long as we are within the area of feasible solutions.   Obviously we shall have to stop only when we have hit either a corner point of the convex polygon of Figure 2.5 or one of its boundary lines.*   In either case we would have found our optimum solution(s).

* A convex polygon consists of a set of points having the property that the segment joining any two points in the set is *entirely* in the convex set.   There is a mathematical theorem which states: "The points which are simultaneous solutions of a system of inequalities of the $\leq$ type form a polygonal convex set."   For purposes of visualization refer to Figures 2.6 and 2.7; the former represents a polygonal convex set, whereas

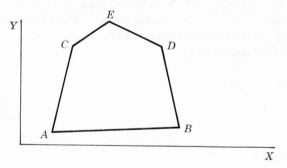

Figure 2.6   Convex polygon.

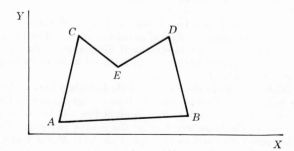

Figure 2.7   Nonconvex polygon.

Finding the Optimum Solution(s)

In a profit-maximization problem involving only two competing candidates, the isoprofit line farthest from the origin but still within the feasible-solution area is used to determine the optimum solution. Two cases can arise. First, this isoprofit line may be coincident with one of the boundary lines of the convex polygon. If this is the case, all points on that boundary line which is coincident with the isoprofit line are feasible as well as optimum solutions.* Secondly, the isoprofit line may not coincide with one of the boundary lines of the convex polygon. If this is the case, one of the corner points of the convex polygon provides the optimum and unique solution.

In our case, we observe that the isoprofit line farthest from the origin and still within the feasible solution area passes through the point H. Hence, point H provides the optimum solution from among an infinite number of solutions represented by the shaded area of Figure 2.5.

The coordinates of the point H can be determined either directly from the graph or by a simultaneous solution of the two lines intersecting at point H. The equations of these lines, representing the cutting and folding capacities, are already known to us. The determination of the coordinates of point H is illustrated below.

For the cutting department

$$10.7X + 5.0Y = 2,705$$

or

$$Y = 541 - 2.14X \tag{4}†$$

---

the latter does not. Note that if points C and D in Figure 2.7 are joined, the definition of a convex polygon is seen to be violated.

Although the illustration is in a two-dimensional space, the concept of convex set is general and can be extended to any n-dimensional space. Furthermore, since all linear-programming problems contain structural constraints of the $\leq$ type (or constraints which can be converted to the $\leq$ type), the solutions to linear-programming problems form a convex set.

* In a two-candidate case, if the slope of any of the boundary lines is the same as that of any isoprofit line, we can conclude that the linear-programming problem has many optimum solutions. This means that any isoprofit line will be either parallel to or coincident with one of the boundary lines of the convex polygon. The reader can verify that if the profit contributions per unit of products X and Y are, respectively, \$10.0 and \$4.68, all points on the line HL (see Figure 2.5) will be optimum solutions.

† A general equation of a straight line is always of the form $Y = a + bX$, where a is the Y intercept and b is the slope.

For the folding department

$$5.4X + 10.0Y = 2,210$$

or

$$Y = 221 - 0.54X \tag{5}$$

Equating (4) and (5), we obtain

$$541 - 2.14X = 221 - 0.54X$$

or

$$X = 200 \tag{6}$$

Substituting (6) in (4),

$$Y = 541 - 2.14(200)$$
$$= 541 - 428 = 113$$

Hence, the optimum solution is to produce 200 units of X (product A) and 113 units of Y (product B). The profit for this program is

$$10(200) + 15(113) = \$3,695$$

Substituting $X = 200$ and $Y = 113$ in the inequalities (1) to (3), we note that this program fully utilizes the capacities of the cutting and folding departments but leaves 192 units of the packaging-department capacity unused. This can also be observed by examining Figure 2.5, in which the packaging-capacity line is far above the shaded area.

2.4 PROCEDURE SUMMARY FOR THE GRAPHICAL METHOD (A TWO-DIMENSIONAL MAXIMIZATION PROBLEM)

Step 1

Transform the given technical specifications of the problem into inequalities, and make a precise statement of the objective (profit) function.

Step 2

Graph each inequality in its limiting case to obtain a region of possible solutions. In conjunction with the nonnegativity constraints, this will yield a convex polygon containing all feasible solutions.

Step 3

By choosing a convenient profit figure, draw an isoprofit line so that it falls within the shaded area.

Step 4

Move this isoprofit line parallel to itself and farther from the origin* until an optimum solution is determined.

The graphical method can also be used for solving a minimization problem. Of the four steps listed, only the last two need be revised to obtain a complete procedure for solving a two-dimensional minimization problem. In step 3 a convenient cost figure, rather than a profit figure, is chosen, and a corresponding isocost line is drawn. In step 4, the isocost line is moved parallel to itself but closer to the origin† until an optimum solution is determined.

2.5 THE GRAPHICAL METHOD (A THREE-DIMENSIONAL CASE)

The data of Table 2.1 are reproduced in Table 2.3. In so far as we now have three competing candidates, we are concerned with a three-dimensional problem.

The data can be expressed as inequalities, one for each resource. Thus, we have

For the cutting department:

$$10.7X + 5.0Y + 2.0Z \leq 2,705 \qquad (7)‡$$

---

* We assume, of course, that each of the two coefficients in the objective (profit) function is positive. Let us consider two other cases: (1) both of the coefficients in the objective function are negative (e.g., maximize $-10X - 15Y$) and (2) one of the coefficients in the objective function is negative while the other is positive (e.g., maximize $-10X + 15Y$). If, in a maximization problem, both the coefficients in the objective function are negative (as in case 1), it is obvious that the level of each activity should be reduced to zero. In other words, neither X nor Y should be produced. If one of the coefficients in the objective function is positive while the other is negative (as in case 2), it is again obvious that the product with the negative coefficient (product X) should not be produced at all, whereas as much as possible of the product with the positive coefficient (product Y) should be produced within the given constraints.

† The reader should draw a parallel between the assumption implied in this statement and the previous footnote.

‡ The variables X, Y, and Z represent, respectively, the quantities of products A, B, and C to be produced.

Table 2.3   *Process Time by Size and Department*

Department	Size			Constraint per time period
	A	B	C	
Cutting...........................	10.7	5.0	2.0	2,705
Folding...........................	5.4	10.0	4.0	2,210
Packaging........................	0.7	1.0	2.0	445
Profit contribution per unit..........	$10	$15	$20	

For the folding department:

$$5.4X + 10.0Y + 4.0Z \leq 2{,}210 \tag{8}$$

For the packaging department:

$$0.7X + 1.0Y + 2.0Z \leq 445 \tag{9}$$

The objective function to be maximized is

$$10X + 15Y + 20Z$$

Since we have three unknowns (X, Y, Z) in this problem, it must be graphed in a three-dimensional space. As before, we impose the non-negativity constraints

$$X \geq 0 \qquad Y \geq 0 \qquad Z \geq 0$$

We now proceed to graph each inequality in its limiting case. That is, we consider them as equations. Since these equations contain three unknowns and are linear, each will yield a "plane" when plotted in three dimensions. The plotting procedure is the same as in the case of a two-dimensional problem. For example, this is how we plot the plane representing the *cutting-department capacity* [consider inequality (7)]:

If $Y, Z = 0$	then $X = 252.8$
$X, Z = 0$	$Y = 541$
$X, Y = 0$	$Z = 1{,}352.5$

This gives us the plane ABC in Figure 2.8.

For the folding department [consider inequality (8)],

If $Y, Z = 0$ then $X = 409.2$

    $X, Z = 0$ $Y = 221$

    $X, Y = 0$ $Z = 552.5$

This gives us the plane DEF in Figure 2.8.

For the packaging department [consider inequality (9)],

If $Y, Z = 0$ then $X = 635.7$

    $X, Z = 0$ $Y = 445$

    $X, Y = 0$ $Z = 222.5$

This gives us the plane GHI in Figure 2.8.

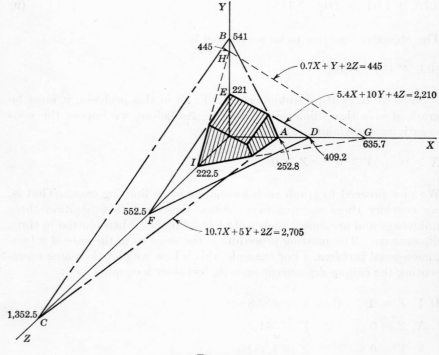

Figure 2.8

The three planes (ABC, DEF, GHI) and the three axes X, Y, Z with $X \geq 0$, $Y \geq 0$, and $Z \geq 0$ form a three-dimensional solid space containing all the feasible solutions to this problem. The next question is: Which particular solution is the optimal solution? To identify this optimal solution, we construct an isoprofit plane.* For example, to obtain an isoprofit plane representing a profit of \$900, we construct a plane whose corners are at $X = 90$, $Y = 60$, and $Z = 45$. We then move this plane away from the origin. In other words, we construct other higher-profit planes parallel to this plane, but still within the region of feasible solutions. The reader can visualize that as we move away from the origin some particular plane will just *touch* the outer surface or most remote corner of the convex polyhedral containing feasible solutions. That point(s) of contact in the three-dimensional space is the optimum solution. As we shall determine later (Chapter 3), the coordinates of this point in our case are

$$X = 200 \qquad Y = 65 \qquad Z = 120$$

This, then, is our optimum solution, resulting in a profit contribution of \$5,375.

Substituting $X = 200$, $Y = 65$, and $Z = 120$ in inequalities (7) to (9) reveals that this program fully utilizes the capacities of all the departments.

2.6 SUMMARY

The graphical method of solving linear-programming problems is limited to cases in which three or less candidates are competing for limited resources. However, the solution of a two- or three-dimensional problem by the graphical method gives us an intuitive insight into the fact that linear-programming problems usually have an infinite number of solutions. Of these, the particular solution(s) which optimizes the objective function is chosen. The other methods, namely, the systematic trial-and-error method, the vector method, and the simplex method, are not limited to problems in three or fewer dimensions. We shall present the systematic trial-and-error method in the next chapter.

---

* This is the same procedure as was used in the two-dimensional plane, where an isoprofit line was drawn to identify the optimal solution.

Systematic Trial-and-Error Method

3.1 INTRODUCTION

Any linear-programming problem in which either the number of competing candidates is limited to 2 or, if there are more than two competing candidates, the number of resources is limited to 2, can always be solved by trial and error.* However, as soon as this restriction is removed, the solution of a linear-programming problem by arbitrary trial and error becomes an almost impossible task. To handle the practical problems of management involving a large number of candidates and resources, therefore, some *systematic* rather than arbitrary trial-and-error method is needed. In other words, we should design some tests or indicators which will guide us as to which way to proceed during our search for an optimum solution(s). The systematic trial-and-error method accomplishes this objective. Before presenting the systematic trial-and-error method, we shall solve the linear-programming problem given in Table 2.2 by arbitrary trial and error. It is, of course, not suggested that the reader consider arbitrary trial and error as a legitimate method of solving linear-programming problems. The *sole* purpose of this detour is to familiarize the reader with some of the terminology used in this book and lay the foundation for understanding the systematic trial-and-error method.

---

* This is so because in linear-programming problems the optimum solution always contains a number of candidates which is equal to or less than the smaller of the numbers of given candidates or resources. For example, in a three-candidate two-resource problem, the optimum solution will not contain more than two products. Thus, the number of possible *combinations* to be tested for their respective values of the objective function is manageable. The rationale of this footnote will become clear to the reader as we progress in this book.

3.2 SOLUTION BY ARBITRARY TRIAL AND ERROR

The data of Table 2.2 are reproduced in Table 3.1 for easy reference.

Table 3.1 *Process Time by Size and Department*

Department	Size		Capacity restriction per time period
	A	*B*	
Cutting..........................	10.7	5.0	2,705
Folding..........................	5.4	10.0	2,210
Packaging.....................	0.7	1.0	445
Profit contribution per unit......	$10	$15	

Transforming the Data into Inequalities

As discussed previously, the problem can be stated as follows:
Maximize $10X + 15Y$ subject to

$$10.7X + 5.0Y \leq 2,705$$

$$5.4X + 10.0Y \leq 2,210$$

$$0.7X + 1.0Y \leq 445$$

and $X \geq 0$, $Y \geq 0$.

Transforming the Inequalities into Equations

Since the above inequalities are of the "less than or equal to" type,* they can be transformed into equations by the addition of nonnegative variables, say S_1, S_2, S_3, etc. Thus, we have

$$10.7X + 5.0Y + S_1 = 2,705$$

$$5.4X + 10.0Y + S_2 = 2,210 \tag{1}$$

$$0.7X + 1.0Y + S_3 = 445$$

---

* See Chapter 7 for the transformation of inequalities of the "greater than or equal to" type.

The variables S_1, S_2, and S_3 are called *slack* variables in the linear-programming literature for they, so to speak, take up the slack and serve to form equalities from the inequalities of a given linear-programming problem. We can also attach a physical interpretation to these slack variables. They can be thought of as "imaginary" products, each requiring for its production 1 unit of capacity from *only* one of the resources and 0 units of capacity from the others, and each yielding a profit of zero. The production of 1 unit of S_1, for example, requires 1 unit of cutting capacity but 0 units of folding capacity and 0 units of packaging capacity.*

The system of Equations (1) has three equations and five unknowns, i.e., more unknowns than equations. This, in general, implies that there are an infinite number of solutions to this set of linear equations. Of course, having solved the same problem by the graphical method, we are already aware of this fact. Since, in general, three equations involving three unknowns can usually be solved for a *unique* solution, two of the five unknowns (X, Y, S_1, S_2, S_3) must be set equal to zero. Theoretically, this would involve solving the system of Equations (1) a total of 10 times.† Some of these solutions may be acceptable solutions while others may not, since they might violate the capacity restrictions or give negative values for either X or Y. But testing all 10 combinations is an extremely lengthy process. Instead, we can reason that in order to create any profit we must institute, in this problem, one of the following types of programs:

1. Produce X alone
2. Produce Y alone
3. Produce feasible combinations of X and Y

We shall now proceed to obtain these programs.

---

* Thus S_1 does not appear in the equations representing folding and packaging capacities. If S_1 were to be included in, say, the equation representing folding capacity, a coefficient of zero would be attached to S_1. The same could be done for S_3 in the folding-capacity equation. Thus, the complete equation for the folding department may read as follows:

$$5.4X + 10.0Y + 0S_1 + 1.0S_2 + 0S_3 = 2{,}705$$

† Combination of five things taken two at a time:

$$C_2^5 = \binom{5}{2} = \frac{5!}{3!2!} = 10$$

Case 1   *Produce X Alone*

This means $Y = 0$, $S_1 = 0$, $S_2 = 0$, and $S_3 = 0$.
The system of Equations (1) then becomes

For cutting capacity:

$$10.7X = 2,705 \quad \text{or} \quad X = 252.8 \checkmark$$

For folding capacity:

$$5.4X = 2,210 \quad \text{or} \quad X = 409.2$$

For packaging capacity:

$$0.7X = 445 \quad \text{or} \quad X = 635.7$$

Since all three resources are used to produce X, we see that in this case the cutting department provides the *limiting* capacity. Hence, the maximum possible production of X is 252.8 units, with a profit of $10(252.8) = \$2,528$.

Case 2   *Produce Y Alone*

This means $X = 0$, $S_1 = 0$, $S_2 = 0$, and $S_3 = 0$.
The system of Equations (1) becomes

For cutting capacity:

$$5Y = 2,705 \quad \text{or} \quad Y = 541$$

For folding capacity:

$$10Y = 2,210 \quad \text{or} \quad Y = 221 \checkmark$$

For packaging capacity:

$$1Y = 445 \quad \text{or} \quad Y = 445$$

Here, the folding capacity is the limiting case. Hence, the maximum possible production of Y is 221 units, yielding a profit of $3,315.

Case 3   Produce X and Y

Naturally, the maximum profit from a program that produces some combination of X and Y will be derived in those cases in which we utilize as much of the given capacities as possible. Thus, we proceed to check those combinations in which either

1. $S_1 = 0$ and $S_2 = 0$, or
2. $S_1 = 0$ and $S_3 = 0$, or
3. $S_2 = 0$ and $S_3 = 0$

1. If, in the system of Equations (1), we let $S_1 = 0$ and $S_2 = 0$, we obtain the following equations:

$$10.7X + 5.0Y = 2{,}705 \tag{2}$$
$$5.4X + 10.0Y = 2{,}210 \tag{3}$$
$$0.7X + 1.0Y + S_3 = 445 \tag{4}$$

Solving (2) and (3) gives $X = 200$ and $Y = 113$.* Substituting these values in (4), we get

$$0.7(200) + 1(113) + S_3 = 445$$

or

$$S_3 = 445 - 140 - 113 = \boxed{192}$$

which means that the solution $X = 200$, $Y = 113$ will necessitate an idle packaging capacity of 192 units. The profit from this program is

$$10(200) + 15(113) = \$3{,}695$$

2. If, in the system of Equations (1), we let $S_1 = 0$ and $S_3 = 0$, then

$$10.7X + 5.0Y = 2{,}705 \tag{5}$$
$$5.4X + 10.0Y + S_2 = 2{,}210 \tag{6}$$
$$0.7X + 1.0Y = 445 \tag{7}$$

---

* Having solved this problem by the graphical method, we know that this is the optimum solution. However, we must test all the combinations listed under case 3.

From (5) and (7),

$$X = 66.67 \qquad Y = 398.33$$

Substituting in (6), we get

$$5.4(66.67) + 10(398.33) + S_2 = 2,210$$

or

$$S_2 = -2,133.32$$

Obviously, this is not an acceptable solution, for it reveals that we are some 2,134 units short in folding capacity.

3. If, in the system of Equations (1), we let $S_2 = 0$ and $S_3 = 0$, then

$$10.7X + 5.0Y + S_1 = 2,705 \tag{8}$$
$$5.4X + 10.0Y = 2,210 \tag{9}$$
$$0.7X + 1.0Y = 445 \tag{10}$$

From (9) and (10), we get

$$X = 1,400 \qquad Y = -535$$

Substituting these values in (8), we obtain

$$S_1 = -9,600$$

This is not an acceptable solution, for it gives a negative value of Y, which has no physical significance.

Hence, after having compared different possible combinations, we come to the conclusion that the most profitable solution in this problem is $X = 200$, $Y = 113$, which is the optimal solution we obtained by the graphical method.

It does not take much imagination to realize that this type of trial-and-error method will get too lengthy and cumbersome as soon as the number of competing candidates and limiting resources increases in a given linear-programming problem. Thus, we must search for a better method, in terms of both savings in computation and getting vital information during the solution. We shall refer to such a method, which is essentially an algebraic method, as the systematic trial-and-error method.

3.3 SOLUTION BY THE SYSTEMATIC TRIAL-AND-ERROR METHOD

In the systematic trial-and-error method, as will be shown in this section, the objective function is used to test the optimality of a given solution. The objective function is modified to yield information as to (1) whether or not the given program can be improved and (2) how to design a new program. Thus, the difficult or even impossible task of determining all possible production combinations (a procedure we followed in Section 3.2) need not be undertaken. Instead, we can design an initial program such that the given constraints are not violated. The initial program can then be tested for optimality by examining the associated objective function. If the test indicates that a better program can be designed, the initial program is revised. The revised program is again tested for optimality, and if further improvement in the objective function is possible, another program is designed. This process is repeated until an optimal solution has been obtained.

We shall illustrate the systematic trial-and-error method by considering the linear-programming problem of Table 2.1. The data are reproduced in Table 3.2 for easy reference. Our task, as before, is to find that optimum "mix" which will yield maximum profit contribution.

Table 3.2   *Process Time by Size and Department*

Department	Size			Capacity per time period
	A	B	C	
Cutting.............................	10.7	5.0	2.0	2,705
Folding.............................	5.4	10.0	4.0	2,210
Packaging...........................	0.7	1.0	2.0	445
Profit per unit of product.............	$10	$15	$20	

Transforming the Data into Inequalities

As previously, our first step is to translate the technical data into inequalities. These inequalities are

For the cutting department:

$$10.7X + 5.0Y + 2.0Z \leq 2,705$$

For the folding department:

$$5.4X + 10.0Y + 4.0Z \leq 2{,}210$$

For the packaging department:

$$0.7X + 1.0Y + 2.0Z \leq 445$$

and $X \geq 0$, $Y \geq 0$, $Z \geq 0$. Our profit function is

$$10X + 15Y + 20Z$$

The interpretation of these inequalities is the same as given previously (see Section 2.3).

Transforming the Inequalities into Equations

Since the above inequalities are of the "less than or equal to" type, they can be transformed into equations by the addition of nonnegative slack variables (each yielding a profit contribution of zero per unit) S_1, S_2, and S_3. Thus, our problem can be stated as follows:

Maximize $10X + 15Y + 20Z + 0S_1 + 0S_2 + 0S_3$ subject to

$$10.7X + 5.0Y + 2.0Z + S_1 = 2{,}705 \tag{11}$$

$$5.4X + 10.0Y + 4.0Z + S_2 = 2{,}210 \tag{12}$$

$$0.7X + 1.0Y + 2.0Z + S_3 = 445 \tag{13}$$

and $X \geq 0$, $Y \geq 0$, $Z \geq 0$. The interpretation of the slack variables is the same as given in Section 3.2.

Designing the Initial Program

Let us now develop a program in which we propose to produce *only* the "imaginary" products S_1, S_2, and S_3.* This means that in Equations (11) to (13) we let $X = 0$, $Y = 0$, and $Z = 0$. Our initial program, therefore, is to produce 2,705 units of S_1, 2,210 units of S_2, and 445 units of S_3.

---

* This will correspond to starting from the origin of our three-dimensional space in Figure 2.8.

Thus, the initial program consists of the following values of the different variables:

$$X = 0 \qquad Y = 0 \qquad Z = 0 \qquad S_1 = 2{,}705 \qquad S_2 = 2{,}210 \qquad S_3 = 445$$

The level of the profit contributions resulting from this program can be determined by substituting the values of the different variables in the objective function. Thus, we have

$$\text{Profit contribution} = 10X + 15Y + 20Z + 0S_1 + 0S_2 + 0S_3$$
$$= 10(0) + 15(0) + 20(0) + 0(2{,}705)$$
$$+ 0(2{,}210) + 0(445)$$
$$= 0$$

The initial program, along with other information, is contained in the following equations. These equations are obtained by a simple rearrangement of Equations (11) to (13).*

$$S_1 = 2{,}705 - 10.7X - 5.0Y - 2.0Z \tag{14}$$

$$S_2 = 2{,}210 - 5.4X - 10.0Y - 4.0Z \tag{15}$$

$$S_3 = 445 - 0.7X - 1.0Y - 2.0Z \tag{16}$$

These equations have physical interpretations. Equation (14), for example, says that if $X = 0$, $Y = 0$, and $Z = 0$, we shall produce 2,705 units of S_1. That is, all the cutting-department capacity will remain idle. Further, Equation (14) reveals that if we want to produce *at this stage*, say, 1 unit of X, then we must be willing to sacrifice 10.7 units of S_1. Similarly, introduction of 1 unit of Y will demand a reduction of S_1 by 5.0 units, and the addition of 1 unit of Z will require reducing S_1 by 2.0 units. In other words, Equation (14) gives us information regarding the *physical ratios of substitution* between X and S_1, Y and S_1, and Z and S_1. These physical ratios of substitution at this stage are nothing but the coefficients of the variables X, Y, Z in the above set of linear equations. However, we should also keep in mind, as the equations reveal, that the introduction of 1 unit of, say, X requires not only the reduction of S_1 by 10.7 units, but the simultaneous reduction of S_2 by 5.4 units and S_3 by 0.7 unit. This, of course, is determined by the manufacturing requirements of the

* Each equation is solved for the product included in a given program. Thus, in this case S_1, S_2, and S_3 are brought to the left-hand sides of the equations.

products as given in Table 3.1. Similar interpretations hold for Equations (15) and (16).

In this problem we note that the addition of a unit of X, Y, or Z makes simultaneous demands for the reduction of S_1, S_2, and S_3. Therefore, the total amount of X, Y, or Z that can, at this stage, be "brought in" the program is limited by the current magnitudes of S_1, S_2, and S_3.

The profit contribution of our initial program is obviously zero.

Revising the Initial Program

In so far as the initial program gives a profit contribution of zero, it can certainly be improved. This improvement is made by designing a new program in such a way that at least *one* of the variables (products) in the present program is replaced by *one* of the variables (products) not in the present program. The replacements are made one at a time. In our example, the variables (products) included in the first program are S_1, S_2, and S_3; the variables (products) external to the first program are X, Y, and Z. Thus, revision of the first program means that one of the variables (products) S_1, S_2, or S_3 must be replaced by either X, Y, or Z. Two questions, in other words, must be answered:

1. Which one of the variables (products) not in the present program should be "brought in" to replace one of the variables (products) currently in the program?
2. What is the maximum amount of the *chosen* variable (product) that can be "brought in"? The answer to this question will also identify the product to be replaced.

In order to determine the particular product to be brought in, we must make a comparison of the cost or profit consequences associated with the introduction of 1 unit of each such product that is currently not included in the program. Then, the variable (product) with the highest net advantage per unit is chosen to be included in the revised program. For example, an examination of our initial program [contained in Equations (14) to (16)] shows that producing 1 unit of Z will require sacrificing 2 units of S_1, 4 units of S_2, and 2 units of S_3. In so far as S_1, S_2, and S_3 have profit contributions of zero per unit, while Z has a profit contribution of $20 per unit, the introduction of Z at this stage would be a desirable course of action. Similar comments can be made in connection with introducing 1 unit of X or Y.

The important question to be asked while revising a given program is: What do we gain by bringing in, say, 1 unit of a particular variable (product) not in the current program, and what do we lose by sacrificing some corresponding* quantities of the variables (products) included in the current program? If the gain is more than the loss, the substitution is desirable; otherwise, not. At this stage of our solution, the net advantage associated with the introduction of 1 unit of Z is \$20, while 1 unit of X or Y yields a net advantage of \$15 or \$10, respectively. We therefore would want to introduce the variable (product) Z in our next program.

Another way to reach the same conclusion is to incorporate the equations representing the present program in the *corresponding* objective function. In this manner, we can immediately identify those variables which can be used for revising the current program to give a net increase in profit. Since this is a maximization problem, these variables will be those which have positive coefficients in the modified objective function. Of these, the variable with the highest net advantage (largest positive coefficient) is chosen to be introduced in the next program.

The above approach for selecting the particular variable to be introduced in the next program is mechanical in nature and quite easy to use. Let us illustrate.

Before designing the initial program, our objective function was

$$\text{Profit contribution} = 10X + 15Y + 20Z + 0S_1 + 0S_2 + 0S_3$$

If we incorporate our present program [represented by Equations (14) to (16)] in the objective function, we obtain

$$
\begin{aligned}
\text{Profit contribution} = {} & 10X + 15Y + 20Z \\
& + 0(2{,}705 - 10.7X - 5Y - 2Z) \\
& + 0(2{,}210 - 5.4X - 10Y - 4Z) \\
& + 0(445 - 0.7X - 1Y - 2Z) \\
= {} & 10X + 15Y + 20Z
\end{aligned}
$$

Since Z has the largest positive coefficient, it is the variable (product) to be first introduced into the solution. At this stage of the solution, therefore, Z is the *key* variable.† This answers our first question as to

---

* The corresponding quantities are determined by the ratios of substitution operating in a *given* program.

† See Chapter 6 for the corresponding concept of *key column*.

which product is to be brought in. To answer the second question, concerning the maximum amount of the chosen product that can be brought in, we reason as follows:

It has already been established that the net advantage associated with the introduction of 1 unit of Z in our next program, at this stage, is \$20. Since this is a profitable course of action, we should continue to bring in Z until one of the currently produced products S_1, S_2, and S_3 (idle capacities) is eliminated from the initial program. The production of Z beyond the level determined in this manner is, of course, not possible, for that would violate some of the nonnegativity constraints; that is, the given "resource" capacities could not support the production of Z beyond the level determined in the above-mentioned manner.

Therefore, to determine the maximum possible level of Z that can be produced in order to improve the initial program, let us examine Equations (14) to (16). These equations, as stated earlier, represent our initial program involving the production of $S_1 = 2,705$ units, $S_2 = 2,210$ units, and $S_3 = 445$ units. Knowing that only Z is to be brought in the solution, we set $X = 0$ and $Y = 0$ in Equations (14) to (16) and increase the value of Z until the left-hand side of *one* of these equations becomes zero. That is, the amount of Z to be brought in is that which will make some S_i (S_1 or S_2 or S_3) the first to become zero. This would mean that the chosen variable (product) Z has completely eliminated one of the variables (S_1, S_2, S_3) in the *current program*. In order to determine which of these variables (S_1, S_2, S_3) will be eliminated from program 1, we set $X = 0$, $Y = 0$ and $S_1 = 0$, $S_2 = 0$, $S_3 = 0$ in Equations (14) to (16). We find that the variable (product) S_3 will be eliminated from the current program and that the maximum amount of Z that can be introduced into the next program is 222.5 units:

From Equation (14):

$$\text{Maximum } Z = \frac{2,705}{2.0} = 1,352.5 \text{ units}$$

From Equation (15):

$$\text{Maximum } Z = \frac{2,210}{4} = 552.5 \text{ units}$$

From Equation (16):

$$\text{Maximum } Z = \frac{445}{2} = 222.5 \text{ units } \checkmark$$

Thus, we have answered our second question: What is the maximum amount of the chosen product that can be brought in?

The above calculations also determine that the variable (product) S_3 will be replaced by Z in the next program. Equation (16), therefore, is what is called the *limiting* or *key* equation.*

Since Z is going to replace S_3, we rearrange Equation (16) so the variable Z is on the left-hand side. Thus,

$$2Z = 445 - 0.7X - 1Y - S_3$$

or

$$Z = 222.5 - 0.35X - 0.5Y - 0.5S_3 \tag{17}$$

This equation can be interpreted as follows. Assuming that $X = 0$, $Y = 0$, and $S_3 = 0$, we shall be producing 222.5 units of Z. Furthermore, this equation gives information, as previously explained, on the ratios of substitution (at this solution stage) between (X, Y, S_3) and Z.

The production of Z, in addition to reducing the production of S_3 to zero, also reduces the amounts of S_1 and S_2 produced. We can get this information in precise terms by substituting Equation (17) in Equations (14) and (15).

Substituting (17) in Equation (14),

$$S_1 = 2{,}705 - 10.7X - 5Y - 2(222.5 - 0.35X - 0.5Y - 0.5S_3)$$
$$= 2{,}260 - 10X - 4Y + S_3 \tag{18}$$

Substituting (17) in Equation (15),

$$S_2 = 2{,}210 - 5.4X - 10Y - 4(222.5 - 0.35X - 0.5Y - 0.5S_3)$$

or

$$S_2 = 1{,}320 - 4X - 8Y + 2S_3 \tag{19}$$

Equation (18) gives us the number of S_1 units remaining (idle capacity of cutting department) as well as the pertinent rates of substitution among the different variables *at this stage*. Equation (19) gives us similar information in connection with the folding department. Our revised program

* See Chapter 6 for the corresponding concept of *key row*.

(program 2) therefore is

$$Z = 222.5 \text{ units} \qquad X = 0 \qquad Y = 0$$
$$S_3 = 0 \qquad\qquad S_1 = 2,260 \qquad S_2 = 1,320$$

Is this an optimal program? As explained earlier, the optimality of a given program is tested by incorporating that program in the corresponding objective function. At any solution stage, this is accomplished by substituting the *rearranged limiting equation* into the corresponding objective function. For example, while revising the initial program, we observed that Equation (16) was the limiting or key equation. This was rearranged to yield Equation (17). Thus, we test the optimality of the current program (program 2) by substituting (17) into the present objective function:

$$\begin{aligned} \text{Profit contribution} &= 10X + 15Y + 20Z \\ &= 10X + 15Y + 20(222.5 - 0.35X - 0.5Y - 0.5S_3) \\ &= 4,450 + 3X + 5Y - 10S_3 \qquad\qquad (20) \end{aligned}$$

The above modified profit function, at this stage, gives us the following information:

Program 2 gives a total profit contribution of $4,450. An additional unit of X, if it can be produced at this stage within the given constraints, will bring a net advantage of $3; a unit of Y will add $5 to the total profit contribution; the addition of 1 unit of S_3 at this stage will subtract $10 from the profit function.

It is important to know why, in Equation (20), the profit contribution of 1 unit of X has become $3, as compared with $10 as given in Table 3.2. The reason for this becomes clear if we keep in mind that Equation (20) represents the modified objective function corresponding to the solution stage given by program 2, which has fully utilized the packaging capacity. Thus, the production of 1 unit of X, at this stage, will mean that 0.7 unit of packaging capacity (required for producing 1 unit of X) must be reallocated from Z. Since each unit of Z requires 2 units of packaging capacity, Z's current level of production will thus be reduced by 0.7/2 unit, which will mean a reduction of $7 (0.7/2 × $20 = $7) in the profit contribution. Thus, the introduction of 1 unit of X, at this stage, results in a gross increase in profit of $10 (attributable to X) but causes a decrease of $7 in the profit contribution of Z. Hence, the introduction of one unit of X, at this stage, will give a *net* profit of $3, which is the coefficient of the variable X in Equation (20).   Similar interpretations can be given to the coefficients of the variables in the objective function at different solution stages.

In so far as its objective function has variables with positive coefficients, a program can be improved.   This being the case in Equation (20), program 2 is not an optimal program.   Furthermore, since Y has the largest positive coefficient in (20), we should introduce Y in the next program.

Program 3 (Revision of Program 2)

According to the guiding rule which we established earlier, we now propose to introduce Y into the solution.   In determining the maximum amount of Y that can be brought in without violating the nonnegativity constraints, we examine Equations (17) to (19), which represent the second program.   Assuming X, Z, S_1, S_2, and S_3 to be zero, we have

From Equation (17):

$$\text{Maximum } Y = \frac{222.5}{0.5} = 445$$

From Equation (18):

$$\text{Maximum } Y = \frac{2,260}{4} = 565$$

From Equation (19):

$$\text{Maximum } Y = \frac{1,320}{8} = 165 \checkmark$$

Hence, Equation (19) provides the limiting case, and the maximum amount of Y that can be brought in is 165 units.   The rationale for this procedure, as the reader will recall, was explained when we revised the initial program to obtain program 2.   Note that while the maximum amount of the chosen variable (product) that can be brought in the solution is being determined, all variables except the "incoming" variable (at this stage, the incoming variable is Y) are given a value of zero in the set of equations representing the current program.

Since Equation (19) is the limiting equation, the introduction of Y must, of course, completely eliminate S_2 from the solution.

Rearranging Equation (19),* we get

$$8Y = 1,320 - 4X + 2S_3 - S_2$$

or

$$Y = 165 - 0.5X + 0.25S_3 - 0.125S_2 \tag{21}$$

The reader, by now, must have learned to appreciate the information that can be obtained from these equations. Equation (21), for example, informs us that if $X = 0$, $S_2 = 0$, and $S_3 = 0$, we shall be producing 165 units of Y. But, what about Z and S_1? To answer this question, we substitute (21) in (17) and then put (21) in (18). Substituting (21) in (17),

$$Z = 222.5 - 0.35X - 0.5(165 - 0.5X + 0.25S_3 - 0.125S_2) - 0.5S_3$$

or

$$Z = 140 - 0.1X + 0.0625S_2 - 0.625S_3 \tag{22}$$

Substituting (21) in (18),

$$S_1 = 2,260 - 10X - 4(165 - 0.5X + 0.25S_3 - 0.125S_2) + S_3$$
$$= 1,600 - 8X + 0.5S_2 \tag{23}$$

Equations (21) to (23) indicate the composition of program 3. The program consists of

$Y = 165$ units $Z = 140$ units $S_1 = 1,600$ units

$X = 0$ $S_2 = 0$ $S_3 = 0$

Is this the optimum program? To answer this question we again derive and examine the corresponding profit function. Substituting (21) in (20),

$$\text{Profit} = 4,450 + 3X + 5Y - 10S_3$$
$$= 4,450 + 3X + 5(165 - 0.5X + 0.25S_3 - 0.125S_2) - 10S_3$$
$$= 5,275 + 0.5X - 8.75S_3 - 0.625S_2 \tag{24}$$

* As explained earlier, the equation to be rearranged is always that which relates the variable (product) to be produced and the product to be completely eliminated from the current program. This, of course, means that the limiting equation is rearranged.

We observe that the present program (program 3) gives a total profit of $5,275. Although we have improved our profit position as compared with the previous program, an optimum program has not been obtained as yet. An examination of the modified profit function [Equation (24)] indicates that we are still left with one positive-coefficient term. Of the three variables (X, S_2, S_3) now represented in the profit function, only X has a positive coefficient. Hence, by bringing X into the solution, total profit can be further increased.

Program 4 (Revision of Program 3)

We have already established, by examining the modified profit function [Equation (24)], that the product X should be brought into the solution. In order to determine the maximum amount of X that can be brought in without violating the nonnegativity constraints, we examine Equations (21) to (23). Assuming Y, Z, S_1, S_2, and S_3 to be zero,

From Equation (21):

$$\text{Maximum } X = \frac{165}{0.5} = 330$$

From Equation (22):

$$\text{Maximum } X = \frac{140}{0.1} = 1,400$$

From Equation (23):

$$\text{Maximum } X = \frac{1,600}{8} = 200 \checkmark$$

Hence, Equation (23) provides the *limiting* case, and the maximum amount of X that can be brought in is 200 units. The introduction of X will reduce the amount of S_1 produced to zero. (Why?)

Rearranging Equation (23), we get

$$8X = 1,600 - S_1 + 0.5S_2$$

or

$$X = 200 - 0.125S_1 + 0.0625S_2 \tag{25}$$

In order to determine the reduction in Y and Z that has to be made by bringing in X, we substitute (25) in (21) and (22).* Substituting (25) in (21),

$$Y = 165 - 0.5(200 - 0.125S_1 + 0.0625S_2) + 0.25S_3 - 0.125S_2$$
$$= 65 + 0.0625S_1 - 0.156S_2 + 0.25S_3 \tag{26}$$

Substituting (25) in (22),

$$Z = 140 - 0.1(200 - 0.125S_1 + 0.0625S_2) + 0.0625S_2 - 0.625S_3$$
$$= 120 + 0.0125S_1 + 0.05625S_2 - 0.625S_3 \tag{27}$$

Equations (25) to (27) give us our latest program: $X = 200$, $Y = 65$, $Z = 120$ and $S_1 = 0$, $S_2 = 0$, $S_3 = 0$.

To determine the total profit produced by this program, and to determine whether any further changes in the program will improve the profit position, we substitute (25) in the corresponding objective function, that is, (24). Hence,

$$\text{Profit} = 5{,}275 + 0.5(200 - 0.125S_1 + 0.0625S_2) - 8.75S_3 - 0.625S_2$$

or

$$\text{Profit} = \$5{,}375 - 0.0625S_1 - 0.593S_2 - 8.75S_3$$

The above modified profit function indicates that the total profit of the present program is \$5,375. In addition, it reveals that no further improvement in the program is possible, for the coefficients of all the variables in the profit function are now negative. An *optimal* solution has therefore been obtained.

---

* Note the particular pattern of the systematic trial-and-error method. Once the incoming variable (product) has been identified, it is for that variable that the *limiting* equation is solved. This solution is inserted into the remaining equations representing a particular program and into the corresponding objective function. This pattern systematically repeats itself from program to program until the optimal solution is determined.

A summary of the various stages of the solution is given in Table 3.3.

Table 3.3

Program	Variables (products) in solution	Variables (products) external to program	Total profit
1	S_1, S_2, S_3	X, Y, Z	0
2	S_1, S_2, Z	X, Y, S_3	\$4,450
3	S_1, Y, Z	X, S_2, S_3	\$5,275
4	X, Y, Z	S_1, S_2, S_3	\$5,375

3.4 PROCEDURE SUMMARY FOR THE SYSTEMATIC TRIAL-AND-ERROR METHOD (MAXIMIZATION CASE)

Step 1   Formulate the Problem

a. Translate the technical specifications of the problem into inequalities and make a precise statement of the objective (profit) function.

b. Convert the inequalities into equalities by the addition of nonnegative slack variables.   Attach a per-unit profit of zero to each slack variable or "imaginary" product.

Step 2   Design an Initial Program

Design an initial program so that only the "imaginary" products are being produced, that is, only the slack variables are included in the solution. Represent the initial program by arranging the equations of step 1 such that the products being produced are on the left-hand sides.

Step 3   Revise the Current Program

a. Identify the incoming variable.   In so far as the initial program consists of only the imaginary products (slack variables), its profit contribution is zero.   Thus, to improve the initial program, the variable with the largest positive coefficient is chosen as the incoming variable.   For programs other than the initial program, the incoming variable is identified by step 4*b*.

b. Determine the maximum quantity of the incoming variable.   From the equations representing the current program, determine the limiting or key equation which will indicate the maximum quantity of the chosen product (variable) that can be introduced into the solution without violating the given constraints.

c. *Obtain equations representing the new program.* Solve for the incoming variable from the limiting equation, and substitute it in the remaining equations of the current program. The new equations represent the revised program.

Step 4   Test for Optimality

a. Substitute the limiting equation (from step 3c) in the corresponding objective function. If there is no positive-coefficient term in the modified objective function, the problem is solved.

b. Otherwise, the program should be revised by bringing in the largest positive-coefficient variable included in the modified objective function.

Step 5

Repeat steps 3b and c and 4 until an optimal program has been designed. An optimal solution has been found when all coefficients in the revised objective function (step 4a) are negative.

3.5 SYSTEMATIC TRIAL-AND-ERROR METHOD (MINIMIZATION CASE)

The procedure for solving a linear-programming problem in which the objective function is to be minimized rather than maximized is exactly the same as that given above except that (1) the variable with the most negative coefficient is chosen for purposes of revising the successive programs and (2) an optimal solution has been found when all coefficients in the revised objective function have become positive.

The reader is encouraged to design a minimization problem and solve it by applying the systematic trial-and-error method.

In conclusion, we may observe that the systematic trial-and-error method, although quite general, becomes rather cumbersome if a given problem involves a large number of variables. However, familiarity with this method will give the reader insight which will be very useful in mastering the simplex method to be discussed in Chapters 6 and 7.

Matrices and Vectors

4.1 INTRODUCTION

Matrix algebra is extremely useful in solving a set of linear equations. As such, any linear-programming problem, if it has a solution, can be solved with the help of matrix algebra.* Furthermore, the algorithm (a systematic procedure) of the simplex method is based on the concepts of matrices, vectors, determinants, and inversion of matrices. It is, therefore, very essential that we become familiar with matrices and vectors and some of their basic properties. This chapter is devoted to accomplishing this task.

4.2 MATRICES

Definition and Notation

A matrix is a rectangular array of ordered numbers. The purpose of a matrix is to convey information in a concise fashion and lend ease of mathematical manipulation. Although a given matrix does not imply any mathematical operations, matrix algebra is a powerful tool for solving a system of linear equations.

Given below are some examples of matrices:

$$A = \begin{bmatrix} 2 & -1 \\ 1 & 3 \end{bmatrix} \qquad B = \begin{bmatrix} 2 & 1 & 4 \\ -3 & 2 & 1 \end{bmatrix} \qquad C = \begin{bmatrix} 10.7 & 5 & 2 \\ 5.4 & 10 & 4 \\ 0.7 & 1 & 2 \end{bmatrix}$$

* The reader will recall that the systematic trial-and-error method of solving a linear-programming problem was really an exercise in the solution of a set of linear equations embodying the given constraints.

The closed brackets of the form [   ] are used to denote a matrix.* Any matrix in which the number of rows equals the number of columns is called a *square* matrix.   Thus, matrices A and C above are square matrices, while B is a rectangular matrix having two rows and three columns.

In general, a matrix A having m rows and n columns is written as

$$A = \begin{bmatrix} a_{11} & a_{12} & \cdots & a_{1n} \\ a_{21} & a_{22} & \cdots & a_{2n} \\ \cdots & \cdots & \cdots & \cdots \\ a_{m1} & a_{m2} & \cdots & a_{mn} \end{bmatrix}$$

In this chapter, we shall use capital letters such as A, B, and C to denote the entire matrix, and small letters with proper subscripts a_{11}, a_{12}, etc., to denote the numbers within the matrix.

The Dimension of a Matrix

The number of rows and columns in a given matrix determines the *dimension* or *order* of the matrix.   For example, consider

$$D = \begin{bmatrix} 2 & 1 \\ -1 & 4 \end{bmatrix} \quad \text{and} \quad E = \begin{bmatrix} -3 & 2 & 1 \\ 2 & 1 & 3 \end{bmatrix}$$

Matrix D is a 2×2 (two by two) matrix, whereas matrix E is a 2×3 matrix.   When specifying the order or dimension of a matrix, the first number always refers to the rows of the matrix, and the second number to the columns of the matrix.   For example, the dimension of a matrix with m rows and n columns is $m \times n$.   Rows of a matrix are numbered from top to bottom; columns are numbered from left to right.

Elements of a Matrix

The different numbers within a matrix are referred to as the *elements* of the matrix.   For example, matrix $A = \begin{bmatrix} 2 & -1 \\ 1 & 3 \end{bmatrix}$ has four elements: 2,

---

* Sometimes matrices are denoted by brackets of the form (   ) or by double vertical lines ‖   ‖.

−1, 1, and 3.   The general form of a 2×2 matrix is given below:

$$A = \begin{bmatrix} a_{11} & a_{12} \\ a_{21} & a_{22} \end{bmatrix}$$

The elements of the matrix are denoted by double subscripts.   In the element a_{12}, the first subscript refers to the row, and the second subscript refers to the column.   The double subscripts give us the *address* of the element, indicating the specific row and column in which the element may be found.   For example, element a_{22} may be found in the second row and second column.   In general, the element a_{ij} is located in the ith row and jth column.*

Real Matrix

If all the elements of a given matrix are real numbers, the matrix is called a *real matrix*.†

Some Special Matrices

Identity Matrix

The identity matrix (sometimes called the *unit* matrix) is a square matrix and is denoted by I.   It is characterized by the fact that all elements on its main diagonal (the diagonal going from the "northwest" corner to the "southeast" corner) are 1s, whereas all other elements are zero. Given below are two different identity matrices:

$$I = \begin{bmatrix} 1 & 0 \\ 0 & 1 \end{bmatrix} \qquad I = \begin{bmatrix} 1 & 0 & 0 \\ 0 & 1 & 0 \\ 0 & 0 & 1 \end{bmatrix}$$

In other words, the identity matrix may be defined as follows:

$$I = [a_{ij}] \qquad \text{where } a_{ij} = \begin{cases} 1 & \text{when } i = j \\ 0 & \text{when } i \neq j \end{cases}$$

* For the sake of uniformity, i's will refer to rows and j's to columns throughout this book.

† Although the elements of a matrix may be complex numbers of the form $a + ib$, in this book we shall be dealing with real matrices only.

The role of the identity matrix in matrix algebra is very similar to that played by the number 1 in ordinary algebra. Provided they are compatible for multiplication, an identity matrix multiplied by any matrix gives the same matrix.* That is,

$$IA = A$$

and

$$AI = A$$

The Zero Matrix

The zero matrix is a matrix in which all elements are zero. It is denoted as 0. Given below are three examples of zero matrices:

$$0 = \begin{bmatrix} 0 & 0 & 0 \\ 0 & 0 & 0 \\ 0 & 0 & 0 \end{bmatrix} \qquad 0 = \begin{bmatrix} 0 & 0 \\ 0 & 0 \end{bmatrix} \qquad 0 = \begin{bmatrix} 0 & 0 & 0 \\ 0 & 0 & 0 \end{bmatrix}$$

The role of the zero matrix in matrix algebra is very similar to that of zero in ordinary algebra. Provided they are compatible for multiplication, the product of any matrix and the zero matrix is a zero matrix. That is,

$$A0 = 0$$

Transpose of a Matrix

Associated with every $n \times n$ matrix A is another matrix A^T whose rows are the columns of the given matrix A, in exactly the same order. In other words, the first row of A becomes the first column in the derived matrix A^T, the second row becomes the second column, and so on. This derived matrix is called the *transpose* of A and is usually denoted by A^T. Obviously, the transpose of the transpose matrix is the original matrix.

---

* As will be explained later, two given matrices are compatible for multiplication only when the number of columns in the lead matrix equals the number of rows in the lag matrix.

Example

Let

$$A = \begin{bmatrix} 2 & 4 & -3 \\ 1 & 2 & 6 \\ 0 & 1 & 5 \end{bmatrix}$$

Then

$$A^T = \begin{bmatrix} 2 & 1 & 0 \\ 4 & 2 & 1 \\ -3 & 6 & 5 \end{bmatrix}$$

and

$$[A^T]^T = A$$

4.3   VECTORS

Definition

When considered as special cases of a matrix, two types of vectors can be identified: (1) row vectors and (2) column vectors.   A *row vector* is an array of numbers written in a row.   Given below are some examples of row vectors:

$$V_1 = [2 \quad 4 \quad -3] \qquad V_2 = [1 \quad 2 \quad 6] \qquad V_3 = [0 \quad 1 \quad 5]$$

Since it is a special case of a matrix, we can say that V_1 is a 1×3 matrix. Thus, in general, a row vector is a $1 \times n$ matrix, where $n = 1, 2, 3, \ldots$. A *column vector* is an array of numbers written in a column.   Given below are examples of column vectors:

$$U_1 = \begin{bmatrix} 2 \\ 1 \\ 0 \end{bmatrix} \qquad U_2 = \begin{bmatrix} 4 \\ 2 \\ 1 \end{bmatrix} \qquad U_3 = \begin{bmatrix} -3 \\ 6 \\ 5 \end{bmatrix}$$

Since it is a special case of a matrix, we can say that U_1 is a 3×1 matrix.   Thus, in general, a column vector is an $m \times 1$ matrix, where $m = 1, 2, 3, \ldots$.   The numbers in a vector are referred to as the *ele-*

ments or *components* of the vector.   For example, the column vector U_1 has three components, namely, 2, 1, and 0.

Unit Vector

A unit vector is a vector in which one element has the value 1 while the rest of the elements are zeros.   Here are some examples of unit vectors:

$$U_1 = \begin{bmatrix} 1 \\ 0 \\ 0 \end{bmatrix} \qquad U_2 = \begin{bmatrix} 0 \\ 1 \\ 0 \end{bmatrix} \qquad V_1 = [1 \quad 0 \quad 0] \qquad V_2 = [0 \quad 1 \quad 0]$$

Zero Vector

A zero vector is a vector in which all the elements are zero.   Given below are a 1×3 zero row vector and a 3×1 zero column vector:

$$0 = [0 \quad 0 \quad 0]$$

$$0 = \begin{bmatrix} 0 \\ 0 \\ 0 \end{bmatrix}$$

Transpose of a Vector

Consider a $1 \times m$ row vector:

$$V = [a_1 \quad a_2 \quad \cdots \quad a_m]$$

If we write this vector vertically with the same elements in exactly the same order, we obtain the *transpose* of V:

$$V^T = U = \begin{bmatrix} a_1 \\ a_2 \\ \cdot \\ \cdot \\ \cdot \\ a_m \end{bmatrix}$$

The column vector U, then, is the transpose of the row vector V. Obviously, the given row vector V is the transpose of the column vector U. That is,

$$U^T = V \quad \text{or} \quad [V^T]^T = V$$

Graphical Representation of Vectors

A given vector can be represented graphically if it has less than four components. Consider, for example, a vector V_1 consisting of a single number, say 5. This vector can be represented as in Figure 4.1. Similarly,

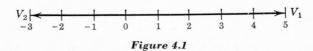

Figure 4.1

a vector $V_2 = -3$ can be represented in a single dimension (see Figure 4.1). In this fashion, we think of vectors as having magnitude as well as *direction*.*

A vector having two elements can be given a graphical interpretation. For example, the vector $V_3 = [4 \quad 2]$ can be graphed as shown in Figure 4.2. We note that, whereas a one-component vector can be graphed in a single dimension, it takes two-dimensional space to graph a two-component vector. By the same token, a three-component vector, say $V_4 = [2 \quad 1 \quad 4]$, can be represented in three-dimensional space (see Figure 4.3).

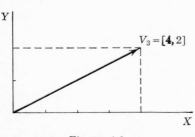

Figure 4.2

In general, then, it takes an n-dimensional space to represent an n-component vector. Evidently, we are limited by our inability to graph a space having more than three dimensions. However, the concept of correspondence between the number of components in a vector and the number of dimensions required to represent it is very important.

---

* A scalar, which is a number, is distinguished from a vector by the fact that a scalar possesses only magnitude.

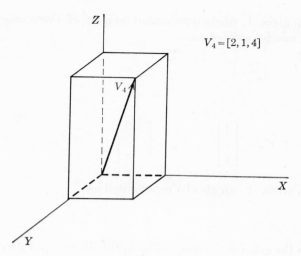

$V_4 = [2, 1, 4]$

Figure 4.3

Vector Notation of a Matrix

If we consider vectors as special cases of a matrix, a given matrix can always be represented as a set of row or column vectors. For example, let A be a 3×3 matrix as given below:

$$A = \begin{bmatrix} 2 & 4 & -3 \\ 1 & 2 & 6 \\ 0 & 1 & 5 \end{bmatrix}$$

The matrix A, when represented as a set of three row vectors V_1, V_2, V_3, can be written as

$$A = \begin{bmatrix} V_1 \\ V_2 \\ V_3 \end{bmatrix}$$

where

$$V_1 = \begin{bmatrix} 2 & 4 & -3 \end{bmatrix}$$
$$V_2 = \begin{bmatrix} 1 & 2 & 6 \end{bmatrix}$$
$$V_3 = \begin{bmatrix} 0 & 1 & 5 \end{bmatrix}$$

The same matrix A, when represented as a set of three column vectors U_1, U_2, U_3, can be written as

$$A = [U_1 \quad U_2 \quad U_3]$$

where

$$U_1 = \begin{bmatrix} 2 \\ 1 \\ 0 \end{bmatrix} \qquad U_2 = \begin{bmatrix} 4 \\ 2 \\ 1 \end{bmatrix} \qquad U_3 = \begin{bmatrix} -3 \\ 6 \\ 5 \end{bmatrix}$$

The same matrix A can also be represented as

$$A = [a_{ij}]$$

where a_{ij} is the general element falling in the ith row and jth column. In order to specify the dimensions of this matrix, the information on the number of rows and columns is given as follows:

$$i = 1, 2, 3$$
$$j = 1, 2, 3$$

4.4   BASIC CONCEPTS AND OPERATIONS CONCERNING MATRICES AND VECTORS

Equality of Matrices

Two matrices are equal if and only if (1) their order is the same and (2) their corresponding elements are equal to each other. Thus, if

$$A = \begin{bmatrix} 2 & 1 & 0 \\ -1 & 3 & 1 \\ 4 & 2 & -1 \end{bmatrix}$$

$$B = \begin{bmatrix} 2 & 1 & 0 \\ -1 & 3 & 1 \\ 4 & 2 & -1 \end{bmatrix}$$

$$C = \begin{bmatrix} 2 & 1 & 0 \\ -1 & 2 & 1 \\ 4 & 2 & -1 \end{bmatrix}$$

then $A = B$, but $A \neq C$, and $B \neq C$. In general, if A and B have the same dimensions, and $a_{ij} = b_{ij}$ for all i and j, then $A = B$.

Equality of Vectors

Two given vectors are said to be equal if and only if (1) they are the same type of vectors and (2) their corresponding elements are equal to each other. Thus, if

$$U = [2 \quad 1 \quad 0] \qquad V = [2 \quad 1 \quad 0] \qquad X = \begin{bmatrix} 2 \\ 1 \\ 0 \end{bmatrix}$$

$$W = [2 \quad -1 \quad 0] \qquad Y = \begin{bmatrix} 2 \\ 1 \\ 1 \end{bmatrix} \qquad Z = \begin{bmatrix} 2 \\ 1 \\ 0 \end{bmatrix}$$

then

$$U = V \qquad U \neq W$$
$$X = Z \qquad X \neq Y$$
$$U \neq X \qquad U \neq Y$$

Addition of Matrices

Two given matrices A and B can be added *only* if they have the same dimensions. Once it is established that the numbers of rows and columns of the two matrices are identical, their respective elements are added together. For this reason, matrix addition is known as *elementwise addition*.

Example

Let

$$A = \begin{bmatrix} 2 & 3 & 4 \\ 1 & 0 & 6 \end{bmatrix} \qquad B = \begin{bmatrix} -1 & 2 & 1 \\ 0 & 3 & 2 \end{bmatrix}$$

Then

$$A + B = \begin{bmatrix} 2 & 3 & 4 \\ 1 & 0 & 6 \end{bmatrix} + \begin{bmatrix} -1 & 2 & 1 \\ 0 & 3 & 2 \end{bmatrix} = \begin{bmatrix} 1 & 5 & 5 \\ 1 & 3 & 8 \end{bmatrix}$$

Two things must be observed in matrix addition. First, the matrix representing the sum of A and B has the same dimension as A and B. Second, the *order* of addition is not important, for if $A + B$ equals C, then $B + A$ also equals C.

In general, if

$$A = \begin{bmatrix} a_{11} & a_{12} & \cdots & a_{1n} \\ a_{21} & a_{22} & \cdots & a_{2n} \\ \cdots & \cdots & \cdots & \cdots \\ a_{m1} & a_{m2} & \cdots & a_{mn} \end{bmatrix} = [a_{ij}] \qquad \begin{aligned} i &= 1, 2, \ldots, m \\ j &= 1, 2, \ldots, n \end{aligned}$$

and

$$B = \begin{bmatrix} b_{11} & b_{12} & \cdots & b_{1n} \\ b_{21} & b_{22} & \cdots & b_{2n} \\ \cdots & \cdots & \cdots & \cdots \\ b_{m1} & b_{m2} & \cdots & b_{mn} \end{bmatrix} = [b_{ij}] \qquad \begin{aligned} i &= 1, 2, \ldots, m \\ j &= 1, 2, \ldots, n \end{aligned}$$

then

$$A + B = B + A = C$$

That is,

$$[a_{ij} + b_{ij}] = [b_{ij} + a_{ij}] = [c_{ij}]$$

where

$$C = \begin{bmatrix} a_{11} + b_{11} & a_{12} + b_{12} & \cdots & & \cdots & a_{1n} + b_{1n} \\ a_{21} + b_{21} & a_{22} + b_{22} & \cdots & & \cdots & \cdots \\ \cdots & \cdots & \cdots & \cdots & \cdots & \cdots \\ \cdots & & \cdots & a_{ij} + b_{ij} & \cdots & \\ \cdots & \cdots & \cdots & \cdots & \cdots & \cdots \\ a_{m1} + b_{m1} & \cdots & & \cdots & \cdots & a_{mn} + b_{mn} \end{bmatrix}$$

As we noted above, the order of addition is not important in matrix addition. We therefore say that matrix addition obeys the *commutative law of addition.* In this sense, matrix addition is similar to the addition of numbers in ordinary algebra.

Addition of Vectors

The addition of vectors, as in the case of matrices, is defined only if both vectors have the same dimensions. This implies that two given vectors can be added only if (1) they are the same type of vectors and (2) they have the same number of elements.

Example

Let

$$V_1 = [2 \quad 3 \quad 4] \qquad V_2 = [-1 \quad 2 \quad 1]$$

Then

$$V_1 + V_2 = [1 \quad 5 \quad 5]$$

Let

$$U_1 = \begin{bmatrix} 2 \\ 1 \end{bmatrix} \qquad U_2 = \begin{bmatrix} -1 \\ 0 \end{bmatrix}$$

Then

$$U_1 + U_2 = \begin{bmatrix} 1 \\ 1 \end{bmatrix}$$

As in the case of matrices, the reader will note that the order of addition of vectors is not important.

Graphic Representation of the Addition of Vectors

Let us consider two row vectors $V_1 = [2 \quad 1]$ and $V_2 = [1 \quad 3]$. Their sum is $V_1 + V_2 = [3 \quad 4]$. This type of addition can be represented graphically as shown in Figure 4.4. The vector V_3 has been obtained by adding the vectors V_1 and V_2. Two things should be observed: (1) Both V_1 and V_2 are two-component vectors and hence need a two-dimensional space to be represented graphically. (2) The vector V_3 lies in the same two-dimensional space. Note that the vector V_3 is the diagonal, passing through the origin, of the parallelogram formed by the vectors V_1 and V_2.

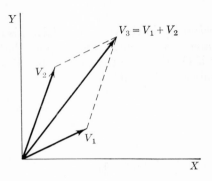

Figure 4.4

Scalar Multiplication and Linear Combination

The simultaneous multiplication of all the elements of a given matrix by a real number is called *scalar multiplication*. Suppose that an $m \times n$ matrix $A = [a_{ij}]$ is to be multiplied by a scalar k. Then

$$Ak = kA = [ka_{ij}]^*$$

Example

Let

$$A = \begin{bmatrix} -2 & 3 & 1 \\ 0 & 2 & 4 \\ 3 & -5 & 1 \end{bmatrix}$$

Then

$$2A = \begin{bmatrix} -4 & 6 & 2 \\ 0 & 4 & 8 \\ 6 & -10 & 2 \end{bmatrix}$$

---

*Since the order of scalar multiplication is not important, we can say that scalar multiplication obeys the commutative law of multiplication.

Similarly, if we multiply all the elements of a given vector by some scalar, we get a scalar multiple of the given vector. In the case of vectors, scalar multiplication can be given a graphical interpretation (see Figure 4.5). Consider, for example, a vector $V_1 = [1 \quad 2]$. Then $2V_1 = V_2 = [2 \quad 4]$. The vector V_2 is graphed in Figure 4.5. As shown in the figure, the multiplication of a given vector by a positive scalar having a value greater than 1 has resulted in the elongation of the given vector without changing its direction. Clearly, the multiplication of a given vector by the scalar 1 will leave its magnitude unaltered, whereas multiplication by a positive scalar having a value less

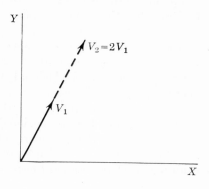

Figure 4.5

than 1 will result in a shortening of the given vector. Thus, if $V_1 = [1 \quad 2]$, then $kV_1 = V_1 = [1 \quad 2]$, where $k = 1$, and $kV_1 = V_2 = [\frac{1}{2} \quad 1]$, where $k = \frac{1}{2}$.

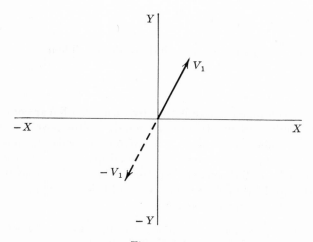

Figure 4.6

The multiplication of a given vector by a positive scalar, therefore, can change the length of the vector without affecting its direction. Let us now multiply V_1 by the scalar -1. Then $-1V_1 = -V_1 = [-1 \quad -2]$. The vector $-V_1$ is graphed in Figure 4.6. The result of multiplying V_1

by the scalar -1 has been, as shown in the figure, to reverse the direction of V_1 without affecting its magnitude.

The illustrations of scalar multiplication (Figures 4.5 and 4.6) serve to indicate that the length as well as the direction of a given vector can be manipulated. Scalar multiplication, in conjunction with the operation of vector addition, gives us a very powerful tool for expressing a given vector in terms of some other vectors. Let us illustrate.

Consider a one-dimensional vector $V_1 = [1]$. All one-dimensional vectors can be expressed as scalar multiples of the vector V_1. For example, if $V_2 = -6$, then $V_2 = -6V_1$. Also, if $V_3 = 8$, then $V_3 = 8V_1$.

Similarly, any two-dimensional vector can be represented by adding scalar multiples of two properly chosen two-dimensional vectors.* Such an addition of scalar multiples of two linearly independent vectors is an example of what is called *linear combination*.

Example

Express $V_3 = [6 \quad 8]$ as a linear combination of two linearly independent vectors.

Let

$$V_1 = [1 \quad 0] \qquad V_2 = [0 \quad 1]$$

Obviously, V_1 and V_2 are linearly independent. Then

$$V_3 = 6V_1 + 8V_2$$

The point to emphasize here is that by forming a linear combination of two linearly independent (two-dimensional) vectors, we can reach any point in a two-dimensional space. By extending this argument, we can say that to reach a point in three-dimensional space, we need three linearly independent (three-dimensional) vectors. Thus, the concept of scalar multiplication and the operation of vector addition are utilized in reaching a point in any n-dimensional space.

* Proper choice in a two-dimensional space means that two *linearly independent* two-dimensional vectors should be chosen. Similarly, in a three-dimensional space, three linearly independent three-dimensional vectors should be chosen. A precise definition of linear independence is given in Section 4.5. For the time being, the reader can consider that a set of vectors is linearly independent if one vector in the set cannot be formed by taking a linear combination of the other vectors in the set. For example, if $V_1 = [1 \quad 0 \quad 0]$, $V_2 = [0 \quad 1 \quad 0]$, and $V_3 = [0 \quad 0 \quad 1]$, then the vectors V_1, V_2, and V_3 are linearly independent. Similarly, if $V_1 = [1 \quad 0]$ and $V_2 = [0 \quad 1]$, then the vectors V_1 and V_2 are again linearly independent.

In particular, a given n-dimensional vector can be represented by a linear combination of n linearly independent n-dimensional vectors.

Subtraction of Matrices

Matrix subtraction is a special case of matrix addition. The rules of matrix subtraction are the same as those obeyed by matrix addition; i.e., the subtraction process is "elementwise subtraction."

Example

Let

$$A = \begin{bmatrix} 2 & 3 & 4 \\ 1 & 0 & 6 \end{bmatrix} \qquad B = \begin{bmatrix} -1 & 2 & 1 \\ 0 & 3 & 2 \end{bmatrix}$$

Then

$$A - B = A + (-B) = \begin{bmatrix} 2 & 3 & 4 \\ 1 & 0 & 6 \end{bmatrix} + \begin{bmatrix} 1 & -2 & -1 \\ 0 & -3 & -2 \end{bmatrix}^*$$

$$= \begin{bmatrix} 3 & 1 & 3 \\ 1 & -3 & 4 \end{bmatrix}$$

In general, if $A = [a_{ij}]$ and $B = [b_{ij}]$, then

$$A - B = A + [-B] = [a_{ij} - b_{ij}] = [c_{ij}] = C$$

Subtraction of Vectors

The subtraction of vectors, as in the case of matrices, is defined only if both vectors have the same dimensions. This implies that a given vector V_2 can be subtracted from another vector V_1 if and only if (1) they are the same type of vectors and (2) they have the same number of elements.

Example

Let

$$V_1 = [2 \quad 3 \quad 4] \qquad V_2 = [-1 \quad 2 \quad 1]$$

---

* Associated with any matrix B is another matrix $-B$ which is obtained by multiplying all the elements of B by -1. This is an example of *scalar multiplication*.

Then

$$V_1 - V_2 = V_1 + (-V_2) = [2 \quad 3 \quad 4] + [1 \quad -2 \quad -1]$$
$$= [3 \quad 1 \quad 3]$$

Multiplication of Vectors

Row Vector × Column Vector

We shall first define the product of a row vector and a column vector, both having the same number of elements.   Let

$$V_1 = [2 \quad 3 \quad 4] \qquad U_1 = \begin{bmatrix} 1 \\ -2 \\ 4 \end{bmatrix} \begin{bmatrix} 2 & 3 & 4 \end{bmatrix}$$

Then

$$V_1 \times U_1 = 2(1) + 3(-2) + 4(4) = [12]$$

In general, the product of a $1 \times n$ row vector by an $n \times 1$ column vector is given by

$$[a_1 \quad a_2 \quad \cdots \quad a_n] \begin{bmatrix} b_1 \\ b_2 \\ \cdot \\ \cdot \\ \cdot \\ b_n \end{bmatrix} = [a_1b_1 + a_2b_2 + \cdots + a_nb_n] = \left[\sum_{i=1}^{n} a_ib_i \right]$$

It is to be observed that the number of elements in the lead vector $[a_1 \quad a_2 \quad \cdots \quad a_n]$ is exactly the same as the number of elements in the lag vector.   If this condition does not exist, the vectors are said to be incompatible, and their multiplication is not defined.   Furthermore, in any multiplication involving a row vector and a column vector, the product is a vector containing a single element, provided the row vector is the lead vector.   This argument can be verified by a check on dimensionality.
   If

$$V_1 = [2 \quad 3 \quad 4] \qquad \text{and} \qquad U_1 = \begin{bmatrix} 1 \\ -2 \\ 4 \end{bmatrix}$$

then

$$V_1U_1 = [12]$$

or

$$V_1 \quad \times \quad U_1 \quad = \quad V_1 U_1$$
Dimensions: $(1 \times 3) \times (3 \times 1) = (1 \times 1)$

Note that the order 1×1 means that we have a single element.

In general, the following dimensional arrangement must hold for compatibility in vector multiplication:

[Lead vector] $\times$ [lag vector] = [product]
Dimensions: $(1 \times n) \quad \times \quad (n \times 1) \quad = \quad (1 \times 1)$

Column Vector × Row Vector*

Let

$$U_1 = \begin{bmatrix} 2 \\ 1 \\ -3 \end{bmatrix} \qquad V_1 = [1 \quad -2 \quad 3]$$

Then

$$U_1 \times V_1 = \begin{bmatrix} 2 \\ 1 \\ -3 \end{bmatrix} [1 \quad -2 \quad 3] = \begin{bmatrix} 2 & -4 & 6 \\ 1 & -2 & 3 \\ -3 & 6 & -9 \end{bmatrix}$$

It is to be observed again that U_1 and V_1 above are dimensionally compatible. However, the product in this case (when the column vector is the lead vector) is a 3×3 matrix.

In general, if the lead vector is an $n \times 1$ column vector and the lag vector is a $1 \times n$ row vector, their product results in an $n \times n$ matrix:

[Lead vector] $\times$ [lag vector] = [product]
Dimensions: $(n \times 1) \quad \times \quad (1 \times n) \quad = \quad (n \times n)$

---

* The multiplication of a row vector by a row vector or of a column vector by a column vector, called a *dot* product, yields a scalar. The dot product is defined as follows:

Let

$$V = [V_1 \quad V_2 \quad \cdots \quad V_n] \qquad X = [X_1 \quad X_2 \quad \cdots \quad X_n]$$

Then

$$V \text{ dot } X = V \cdot X = V_1 X_1 + V_2 X_2 + \cdots + V_n X_n = \sum_{i=1}^{n} V_i X_i$$

Multiplication of Matrices

The definition of multiplication of a row vector by a column vector can easily be extended to cover matrix multiplication. Before considering a general case, let us illustrate matrix multiplication by considering a specific example. Let

$$A = \begin{bmatrix} 2 & 1 & -2 \\ 3 & 2 & 4 \end{bmatrix} \qquad B = \begin{bmatrix} 1 & 2 \\ 0 & 3 \\ -2 & 1 \end{bmatrix}$$

and let $AB = C$. Then

$$A \times B = C = \begin{bmatrix} 2(1) + 1(0) + (-2)(-2) & 2(2) + 1(3) + (-2)(1) \\ 3(1) + 2(0) + 4(-2) & 3(2) + 2(3) + 4(1) \end{bmatrix}$$

or

$$C = \begin{bmatrix} 6 & 5 \\ -5 & 16 \end{bmatrix}$$

The above multiplication consists of the following steps:

1. *Check on compatibility.* Is the number of columns in the lead matrix [A] equal to the number of rows in the lag matrix [B]? If so, the matrices are compatible for multiplication; otherwise, not. In the above case, the multiplication $A \times B$ is compatible.

2. *The operation of multiplication.* The elements of the first row of A, the lead matrix, are multiplied by the corresponding elements of the first column of B, the lag matrix. The product is summed and is placed in the first-row first-column cell of the resultant matrix C. Similarly, the elements of the second row of matrix A are multiplied by the corresponding elements of the first column of matrix B; the product is summed and is placed in the second-row first-column element of the resultant matrix, and so on. The resultant matrix C is a 2×2 matrix.

Check on dimensional compatibility:

	[Lead matrix]	×	[lag matrix]	=	[resultant matrix]
	A	×	B	=	C
Dimensions:	(2×3)	×	(3×2)	=	(2×2)

In general, the product of an $m \times k$ matrix A and a $k \times n$ matrix B is given by

$$AB = \left[\sum_{s=1}^{k} a_{is} b_{sj} \right] \quad \text{for} \begin{cases} i = 1, 2, \ldots, m \\ j = 1, 2, \ldots, n \end{cases}$$

and the dimension of the resultant matrix is $m \times n$:

$$[\text{Lead matrix}] \times [\text{lag matrix}] = [\text{resultant matrix}]$$
Dimensions: $(m \times k)$ $\times$ $(k \times n)$ $=$ $(m \times n)$

Example

Let

$$C = \begin{bmatrix} 4 & -1 & 2 \\ 0 & 2 & 3 \end{bmatrix} \quad D = \begin{bmatrix} 2 & 3 & 1 \\ 0 & 1 & 0 \\ 2 & 1 & 0 \end{bmatrix}$$

Then

$$CD = \begin{bmatrix} 12 & 13 & 4 \\ 6 & 5 & 0 \end{bmatrix}$$

but DC is incompatible.

Example

Obtain the product BA when

$$A = \begin{bmatrix} 2 & 1 & -2 \\ 3 & 2 & 4 \end{bmatrix} \quad B = \begin{bmatrix} 1 & 2 \\ 0 & 3 \\ -2 & 1 \end{bmatrix}$$

$$BA = \begin{bmatrix} 1 & 2 \\ 0 & 3 \\ -2 & 1 \end{bmatrix} \begin{bmatrix} 2 & 1 & -2 \\ 3 & 2 & 4 \end{bmatrix}$$

$$= \begin{bmatrix} 1(2) + 2(3) & 1(1) + 2(2) & 1(-2) + 2(4) \\ 0(2) + 3(3) & 0(1) + 3(2) & 0(-2) + 3(4) \\ -2(2) + 1(3) & -2(1) + 1(2) & -2(-2) + 1(4) \end{bmatrix}$$

$$= \begin{bmatrix} 8 & 5 & 6 \\ 9 & 6 & 12 \\ -1 & 0 & 8 \end{bmatrix}$$

In this example, AB and BA are both compatible, but note that

$$AB \neq BA$$

This leads to the remark that in matrix multiplication *order* is important. Matrix multiplication, therefore, does not obey the commutative law of multiplication.

Comparisons between various operations in ordinary algebra and matrix algebra are given in Table 4.1.

Table 4.1

Law	Ordinary algebra	Matrix algebra
1*a.* *Commutative law of addition* (dealing with the order of operations) *Comment:* Matrix addition obeys the commutative law of addition	$a + b = b + a = c$	$A + B = B + A = C$
1*b.* *Commutative law of multiplication* *Comment:* Matrix multiplication, in general, does not obey the commutative law of multiplication; however, two exceptions must be noted: (1) scalar multiplication and (2) multiplication of a square matrix by the identity matrix I	$a \times b = b \times a$	$AB \neq BA$
2*a.* *Associative law* (order remains the same; deals with the sequence of similar operations within a given order) *Comment:* Matrix addition obeys the associative law of addition	$a + (b + c) = (a + b) + c$	$A + (B + C) = (A + B) + C$
2*b.* *Associative law* *Comment:* Matrix multiplication obeys the associative law of multiplication	$a(bc) = (ab)c$	$A(BC) = (AB)C$
3. *Distributive law* (deals with the sequence of addition and multiplication operations within a given order) *Comment:* Matrix algebra obeys the distributive law	$a(b + c) = ab + ac$ $(d + e)f = df + ef$	$A(B + C) = AB + AC$ $(D + E)F = DF + EF$

Multiplication of a Matrix by a Vector

The multiplication of a matrix by a vector follows the rules of regular matrix multiplication as explained previously.

Example

Let

$$A = \begin{bmatrix} 4 & -1 & 2 \\ 0 & 2 & 3 \end{bmatrix} \quad \text{and} \quad U = \begin{bmatrix} 1 \\ 0 \\ 2 \end{bmatrix}$$

Then

$$AU = \begin{bmatrix} 4 & -1 & 2 \\ 0 & 2 & 3 \end{bmatrix} \begin{bmatrix} 1 \\ 0 \\ 2 \end{bmatrix} = \begin{bmatrix} 8 \\ 6 \end{bmatrix}$$

4.5   LINEAR INDEPENDENCE

A set of vectors $V_1, V_2, \ldots, V_m$ of the same dimension is said to be *linearly dependent* if a set of scalars $k_1, k_2, \ldots, k_m$, not all zero, can be found such that

$$k_1V_1 + k_2V_2 + \cdots + k_mV_m = 0$$

where 0 represents a zero vector.

If the above relationship holds only when all the scalars $k_1, k_2, \ldots, k_m$ are zero, then the set of vectors $V_1, V_2, \ldots, V_m$ is said to be *linearly independent.**

Example

Test the vectors $V_1 = \begin{bmatrix} 2 \\ 3 \end{bmatrix}$ and $V_2 = \begin{bmatrix} 4 \\ 2 \end{bmatrix}$ for linear independence.

Let

$$k_1V_1 + k_2V_2 = 0$$

---

* An extremely easy test for linear independence among a given set of m vectors each of dimension $m \times 1$ (or $1 \times m$) is based on the value of the determinant of the square matrix formed by the vectors. If the determinant is zero, the vectors are linearly dependent. If, on the other hand, the determinant is nonzero, the vectors are linearly independent. As we shall see in later sections, the significance of linear independence lies in the fact that all vectors in the same vector space can be expressed in terms of a set of independent vectors. See Section 4.7, which is devoted to a discussion of determinants.

or

$$k_1 \begin{bmatrix} 2 \\ 3 \end{bmatrix} + k_2 \begin{bmatrix} 4 \\ 2 \end{bmatrix} = \begin{bmatrix} 0 \\ 0 \end{bmatrix}$$

or

$$2k_1 + 4k_2 = 0 \tag{1}$$
$$3k_1 + 2k_2 = 0 \tag{2}$$

Multiply Equation (2) by 2 and subtract Equation (1) from the product; then

$$4k_1 = 0 \qquad \text{or} \qquad k_1 = 0 \tag{3}$$

Substituting (3) in (1), we find

$$k_2 = 0$$

Since the relationship $k_1 V_1 + k_2 V_2 = 0$ holds only when $k_1 = 0$ and $k_2 = 0$, the vectors V_1 and V_2 are linearly independent.

Example

Test the vectors $V_3 = \begin{bmatrix} 3 \\ 2 \end{bmatrix}$ and $V_4 = \begin{bmatrix} 6 \\ 4 \end{bmatrix}$ for linear independence.

Let

$$k_3 V_3 + k_4 V_4 = 0$$

or

$$k_3 \begin{bmatrix} 3 \\ 2 \end{bmatrix} + k_4 \begin{bmatrix} 6 \\ 4 \end{bmatrix} = \begin{bmatrix} 0 \\ 0 \end{bmatrix}$$

Clearly, the above equation is satisfied if we let $k_3 = 1$ and $k_2 = -\frac{1}{2}$. Hence, according to the definition of linear dependence, vectors V_3 and V_4 are linearly dependent.   In other words, the set of vectors V_3, V_4 is not a linearly independent set.

4.6   VECTOR SPACE AND BASIS
FOR A VECTOR SPACE

Definition

A set of vectors forms a *vector space* if (1) the sum of any two vectors in the set is also in the set and (2) all scalar multiples of any vector in the set are also in the set.*

Let us illustrate by considering a set of vectors $V_1 = \begin{bmatrix} 2 \\ 3 \end{bmatrix}$ and $V_2 = \begin{bmatrix} 4 \\ 7 \end{bmatrix}$. Clearly, vectors V_1 and V_2 can be represented in a two-dimensional space (see Figure 4.7). As a matter of fact, any two-dimensional vector can be

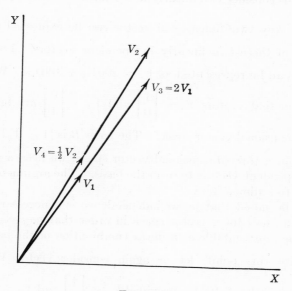

Figure 4.7

represented in a two-dimensional space, provided we consistently use the same geometrical scheme to represent the two components. Now consider $2V_1 = V_3 = \begin{bmatrix} 4 \\ 6 \end{bmatrix}$ and $\frac{1}{2}V_2 = V_4 = \begin{bmatrix} 2 \\ \frac{7}{2} \end{bmatrix}$. Obviously, V_3 and V_4 can also be represented in the XY plane. Indeed, any scalar multiple of V_1, say k_1V_1, or any scalar multiple of V_2, say k_2V_2, can be represented in the two-dimensional space. Furthermore, $V_1 + V_2 = V_5 = \begin{bmatrix} 6 \\ 10 \end{bmatrix}$ can

* Leonard E. Fuller, "Basic Matrix Theory," p. 71, Prentice-Hall, Inc., Englewood Cliffs, N.J., 1962.

also be represented in the same vector space. In general, any linear combination of the scalar multiples of V_1 and V_2, say $k_1 V_1 + k_2 V_2$, is also a vector which can be represented in the same vector space.

Basis for a Vector Space

A *basis* for a vector space is a set of linearly independent vectors such that *any* vector in the vector space can be expressed as a linear combination of this set. The vectors in such a set are called the *basis vectors*. Let us consider two linearly independent vectors $V_1 = \begin{bmatrix} 1 \\ 0 \end{bmatrix}$ and $V_2 = \begin{bmatrix} 0 \\ 1 \end{bmatrix}$.* Any two-dimensional vector can be expressed as a linear combination of these two linearly independent vectors. For example, $V_3 = \begin{bmatrix} 200 \\ 100 \end{bmatrix}$ can be represented as $V_3 = 200 V_1 + 100 V_2$. We come to the conclusion that vectors $V_1 = \begin{bmatrix} 1 \\ 0 \end{bmatrix}$ and $V_2 = \begin{bmatrix} 0 \\ 1 \end{bmatrix}$ are basis vectors for a two-dimensional vector space. The basis B is $[V_1 \quad V_2] = \begin{bmatrix} 1 & 0 \\ 0 & 1 \end{bmatrix}$.

Similarly, in a three-dimensional vector space, we need a set of three linearly independent vectors to form the basis. The argument can easily be extended to n dimensions.

It should be noted that linear independence is a necessary condition for forming a basis for a vector space in order that *any* vector in that space may be representable as a linear combination of the basis vectors. To emphasize this point, let us again consider vector $V_3 = \begin{bmatrix} 200 \\ 100 \end{bmatrix}$. Clearly, $V_3 = 200 V_1 + 100 V_2$, provided $V_1 = \begin{bmatrix} 1 \\ 0 \end{bmatrix}$ and $V_2 = \begin{bmatrix} 0 \\ 1 \end{bmatrix}$, where V_1 and V_2 are linearly independent. But if we consider two linearly dependent vectors, say $V_4 = \begin{bmatrix} 3 \\ 2 \end{bmatrix}$ and $V_5 = \begin{bmatrix} 6 \\ 4 \end{bmatrix}$, then we simply cannot express V_3 as a linear combination of scalar multiples of V_4 and V_5. The reader should verify this statement with a geometrical representation of these vectors.

---

* By applying the definition of linear independence, the reader can check that $V_1 = \begin{bmatrix} 1 \\ 0 \end{bmatrix}$ and $V_2 = \begin{bmatrix} 0 \\ 1 \end{bmatrix}$ are linearly independent.

The concepts of linear independence and basis are very important in linear programming. As the reader will observe in Chapter 6, the first tableau of the simplex method creates a basis, say for an m-dimensional space, by using m linearly independent unit vectors of the type

$$
V_1 = \begin{bmatrix} 1 \\ 0 \\ 0 \\ \cdot \\ \cdot \\ \cdot \\ 0 \end{bmatrix} \qquad
V_2 = \begin{bmatrix} 0 \\ 1 \\ 0 \\ \cdot \\ \cdot \\ \cdot \\ 0 \end{bmatrix} \qquad \cdots \qquad
V_m = \begin{bmatrix} 0 \\ 0 \\ \cdot \\ \cdot \\ \cdot \\ 0 \\ 1 \end{bmatrix}
$$

Further, as the reader will note in Chapter 9, a so-called "degeneracy" occurs in linear-programming problems when an m-dimensional vector is represented as a linear combination of less than m independent vectors.

4.7 REPRESENTING LINEAR EQUATIONS WITH VECTORS AND VICE VERSA

In Section 4.4 we observed that any n-dimensional vector can be represented as a linear combination of n linearly independent vectors. The problem, of course, is to determine the specific combination of given linearly independent vectors with which to form a given vector.

Now that we are familiar with the operations of matrix addition, scalar multiplication, and matrix multiplication, we can solve a specific linear-combination problem for a two-dimensional vector.

Example

Given below are two linearly independent vectors V_1 and V_2:

$$
V_1 = \begin{bmatrix} 2 \\ 3 \end{bmatrix} \qquad V_2 = \begin{bmatrix} 1 \\ 2 \end{bmatrix}
$$

What linear combination of V_1 and V_2 will form $V_3 = \begin{bmatrix} 50 \\ 60 \end{bmatrix}$? In other words, find the scalars k_1 and k_2 such that $k_1 V_1 + k_2 V_2 = V_3$, or

$$
k_1 \begin{bmatrix} 2 \\ 3 \end{bmatrix} + k_2 \begin{bmatrix} 1 \\ 2 \end{bmatrix} = \begin{bmatrix} 50 \\ 60 \end{bmatrix}
$$

By the definition of equality of vectors, we have

$$2k_1 + k_2 = 50 \qquad\qquad (4)$$
$$3k_1 + 2k_2 = 60 \qquad\qquad (5)$$

Multiplying Equation (4) by 2,

$$4k_1 + 2k_2 = 100 \qquad\qquad (6)$$

Subtracting (5) (6) from,

$$k_1 = 40$$

Then, substituting $k_1 = 40$ in (4),

$$2(40) + k_2 = 50$$

or

$$k_2 = -30$$

Thus,

$$V_3 = 40V_1 - 30V_2$$

Note that the specific linear combination above was obtained by transforming vectors into linear equations. By the same token, linear equations can be transformed into vectors. Linear-programming problems involving a set of linear equations are, therefore, amenable to solution by the vector method (see Chapter 5).

4.8 DETERMINANTS

Associated with any square matrix is a number which is called its *determinant*. Consider, for example, a square matrix A:

$$A = \begin{bmatrix} 2 & 1 & 3 \\ -1 & 4 & 1 \\ 3 & 2 & 5 \end{bmatrix}$$

This square matrix has a determinant which, in its unexpanded form, is written as follows:

$$|A| = \begin{vmatrix} 2 & 1 & 3 \\ -1 & 4 & 1 \\ 3 & 2 & 5 \end{vmatrix}$$

The reader should note that the determinant is denoted by single vertical bars, whereas the matrix notation was [ ]. Further, whereas the given matrix A did not imply any mathematical operation, the determinant of this matrix (having exactly the *same* elements in the *same* positions) does imply certain operations. When the determinant appears in the above form, it is said to be in its *unexpanded* form. To evaluate a determinant, we expand it according to certain rules and obtain a single number.

The concept of determinants is very useful. By evaluating the determinant of a given set of linear equations, for example, one can immediately determine whether or not the set has a unique solution. If the given set does not have a unique solution, further tests (also using the concept of determinants) will show that the set either has no solution or has an infinite number of solutions. Furthermore, if a given set of linear equations has a unique solution, its determinant is used to find the values of the unknowns in that set.

How to Evaluate Determinants

Before giving a generalized method of evaluating determinants, we shall show how a 2×2 determinant is evaluated. Consider a matrix

$$A = \begin{bmatrix} 2 & 4 \\ 3 & 7 \end{bmatrix}$$

The determinant of A is

$$|A| = \begin{vmatrix} 2 & 4 \\ 3 & 7 \end{vmatrix} = \begin{vmatrix} 2 & \text{\small main} & 4 \\ 3 & \text{\small diagonal} & 7 \end{vmatrix} = 14 - 12 = 2$$

In other words, *a 2×2 determinant is evaluated by taking the product of its main diagonal elements and subtracting from it the product of the remaining elements.*

In general, if

$$A = \begin{bmatrix} a_{11} & a_{12} \\ a_{21} & a_{22} \end{bmatrix}$$

then

$$|A| = \begin{vmatrix} a_{11} & a_{12} \\ a_{21} & a_{22} \end{vmatrix} = a_{11}a_{22} - a_{21}a_{12}$$

Before discussing the procedure for evaluating determinants of order higher than 2×2, we present the definitions of a minor and a cofactor.

Minor

The determinant of the submatrix formed by deleting *one* row and *one* column from a given square matrix is called a *minor*. Consider, for example, the matrix

$$A = \begin{bmatrix} 2 & 4 & 1 \\ 3 & 7 & 2 \\ 1 & 0 & -4 \end{bmatrix}$$

If we delete the first row and the first column, we get the submatrix

$$A' = \begin{bmatrix} 7 & 2 \\ 0 & -4 \end{bmatrix}$$

The determinant of A' is

$$|A'| = \begin{vmatrix} 7 & 2 \\ 0 & -4 \end{vmatrix}$$

The determinant $|A'|$, as defined above, is the *minor* obtained by deleting the first row and the first column of the given matrix. This minor is denoted by M_{11}. The first subscript of the minor always refers to the row being deleted from the matrix, and the second subscript refers to the column being deleted. Thus,

$$M_{11} = \begin{vmatrix} 7 & 2 \\ 0 & -4 \end{vmatrix} = 7(-4) - (0)(2) = -28$$

Similarly, M_{12} is the minor obtained by deleting the first row and the second column. In our example,

$$M_{12} = \begin{vmatrix} 3 & 2 \\ 1 & -4 \end{vmatrix} = 3(-4) - (1)(2) = -14$$

In general, M_{ij} is the minor obtained by deleting the ith row and the jth column of a given matrix.

Cofactor

Associated with each element of a square matrix is a cofactor. Consider, for example, the following matrix:

$$A = \begin{bmatrix} 2 & 4 & 1 \\ 3 & 7 & 2 \\ 1 & 0 & -4 \end{bmatrix}$$

The cofactor of the element 3 in this matrix (second row, first column) is denoted by C_{21} and is defined as

$$C_{21} = (-1)^{2+1} \begin{vmatrix} 4 & 1 \\ 0 & -4 \end{vmatrix}$$

But we already know that $\begin{vmatrix} 4 & 1 \\ 0 & -4 \end{vmatrix}$ is M_{21}, that is, the minor obtained by deleting the second row and first column of matrix A. The only difference between C_{21} and M_{21}, therefore, is the sign generated by the expression $(-1)^{2+1}$. Indeed, this is exactly what differentiates a cofactor from its minor. A cofactor is a minor with its proper sign, and this sign is determined by the subscripts of the element with which the cofactor is associated. Consider, for example, the following matrix:

$$A = \begin{bmatrix} a_{11} & a_{12} & a_{13} \\ a_{21} & a_{22} & a_{23} \\ a_{31} & a_{32} & a_{33} \end{bmatrix}$$

The cofactor of a_{11} is $(-1)^{1+1} \begin{vmatrix} a_{22} & a_{23} \\ a_{32} & a_{33} \end{vmatrix}$.

In general, the cofactor of any element a_{ij} (where i refers to the row, and j to the column) is denoted by C_{ij}, and its value is given by

$$C_{ij} = (-1)^{i+j} M_{ij}$$

Example

Consider a matrix

$$B = \begin{bmatrix} 1 & 2 & 3 & 2 \\ 4 & -1 & 0 & 1 \\ 0 & 2 & 1 & 3 \\ -3 & 0 & 1 & 2 \end{bmatrix}$$

Find the cofactor C_{24}.

As defined above, the cofactor C_{24} is the minor M_{24} with its proper sign. Hence,

$$C_{24} = (-1)^{2+4} \begin{vmatrix} 1 & 2 & 3 \\ 0 & 2 & 1 \\ -3 & 0 & 1 \end{vmatrix}$$

Note that $(-1)^{i+j}$ is positive when $i + j$ is even, and negative when $i + j$ is odd.

Expansion of a Determinant by Cofactors

The determinant of any square matrix A can be evaluated by expanding it along any row or column as follows: Let

$$A = \begin{bmatrix} a_{11} & a_{12} & a_{13} \\ a_{21} & a_{22} & a_{23} \\ a_{31} & a_{32} & a_{33} \end{bmatrix}$$

The determinant of A is

$$|A| = a_{11}C_{11} + a_{12}C_{12} + a_{13}C_{13} \qquad \text{determinant expanded along first row}$$

or

$$|A| = a_{21}C_{21} + a_{22}C_{22} + a_{23}C_{23} \qquad \text{determinant expanded along second row}$$

Similarly,

$$|A| = a_{11}C_{11} + a_{21}C_{21} + a_{31}C_{31} \qquad \text{determinant expanded along first column}$$

$$|A| = a_{13}C_{13} + a_{23}C_{23} + a_{33}C_{33} \qquad \text{determinant expanded along third column}$$

Note that the subscripts of the element and its cofactor are the same in each term of the expansion. This method of expanding a determinant is general and can be applied to any $n \times n$ determinant. It is sometimes called *Laplace's expansion*.

Example

Consider a matrix

$$A = \begin{bmatrix} 2 & 3 & 2 \\ -1 & 0 & 1 \\ 2 & 1 & 3 \end{bmatrix}$$

The determinant of A is

$$|A| = \begin{vmatrix} 2 & 3 & 2 \\ -1 & 0 & 1 \\ 2 & 1 & 3 \end{vmatrix}$$

Expanding along the second row,

$$|A| = a_{21}C_{21} + a_{22}C_{22} + a_{23}C_{23}$$

$$= (-1)(-1)^{2+1}\begin{vmatrix} 3 & 2 \\ 1 & 3 \end{vmatrix} + 0C_{22} + 1(-1)^{2+3}\begin{vmatrix} 2 & 3 \\ 2 & 1 \end{vmatrix}$$

$$= (9 - 2) + 0 + (-1)(2 - 6)$$

$$= 11$$

Expanding along the second column,

$$|A| = a_{12}C_{12} + a_{22}C_{22} + a_{32}C_{32}$$

$$= 3(-1)^{1+2}\begin{vmatrix} -1 & 1 \\ 2 & 3 \end{vmatrix} + 0C_{22} + 1(-1)^{3+2}\begin{vmatrix} 2 & 2 \\ -1 & 1 \end{vmatrix}$$

$$= 3(-1)(-5) + 0 + (-1)(4)$$

$$= 11$$

4.9   THE COFACTOR MATRIX

Associated with a matrix A is another matrix whose elements are cofactors of the corresponding elements of A.   Such a matrix is called the *cofactor matrix*.   Consider, for example, the matrix

$$A = \begin{bmatrix} a_{11} & a_{12} & a_{13} \\ a_{21} & a_{22} & a_{23} \\ a_{31} & a_{32} & a_{33} \end{bmatrix}$$

The corresponding cofactor matrix is

$$A_{\text{cofactor}} = \begin{bmatrix} C_{11} & C_{12} & C_{13} \\ C_{21} & C_{22} & C_{23} \\ C_{31} & C_{32} & C_{33} \end{bmatrix}$$

4.10   THE ADJOINT MATRIX

The transpose of the cofactor matrix is called the *adjoint matrix*.   Thus, for the above matrix A,

$$A_{\text{adj}} = [A_{\text{cofactor}}]^T$$

or

$$A_{\text{adj}} = \begin{bmatrix} C_{11} & C_{12} & C_{13} \\ C_{21} & C_{22} & C_{23} \\ C_{31} & C_{32} & C_{33} \end{bmatrix}^T = \begin{bmatrix} C_{11} & C_{21} & C_{31} \\ C_{12} & C_{22} & C_{32} \\ C_{13} & C_{23} & C_{33} \end{bmatrix}$$

As we shall see in the next section, the adjoint matrix is very useful in finding the inverse of a given matrix.

4.11   THE INVERSE MATRIX

If, for a given square matrix A, there exists another square matrix B such that $AB = BA = I$ (the identity matrix), then B is said to be the *inverse* of A.   The inverse of A is usually denoted by A^{-1}.   When it exists, A^{-1} plays a role similar to that played by the reciprocal of a given number in ordinary algebra—although it must be noted that not all matrices have inverses.   As a matter of fact, only square matrices with nonzero determinants have inverses.

Example

Let

$$A = \begin{bmatrix} 4 & 0 & 0 \\ 0 & 6 & 2 \\ 2 & 0 & 1 \end{bmatrix} \quad \text{and} \quad B = \begin{bmatrix} \frac{1}{4} & 0 & 0 \\ \frac{1}{6} & \frac{1}{6} & -\frac{1}{3} \\ -\frac{1}{2} & 0 & 1 \end{bmatrix}$$

Then

$$AB = \begin{bmatrix} 4 & 0 & 0 \\ 0 & 6 & 2 \\ 2 & 0 & 1 \end{bmatrix} \begin{bmatrix} \frac{1}{4} & 0 & 0 \\ \frac{1}{6} & \frac{1}{6} & -\frac{1}{3} \\ -\frac{1}{2} & 0 & 1 \end{bmatrix} = \begin{bmatrix} 1 & 0 & 0 \\ 0 & 1 & 0 \\ 0 & 0 & 1 \end{bmatrix}$$

$$BA = \begin{bmatrix} \frac{1}{4} & 0 & 0 \\ \frac{1}{6} & \frac{1}{6} & -\frac{1}{3} \\ -\frac{1}{2} & 0 & 1 \end{bmatrix} \begin{bmatrix} 4 & 0 & 0 \\ 0 & 6 & 2 \\ 2 & 0 & 1 \end{bmatrix} = \begin{bmatrix} 1 & 0 & 0 \\ 0 & 1 & 0 \\ 0 & 0 & 1 \end{bmatrix}$$

That is,

$$AB = BA = I$$

Hence, by definition, $B = A^{-1}$. It may be noted that if a given matrix has an inverse, the inverse is unique.

How to Find an Inverse

We shall illustrate two methods for finding the inverse (if it exists) of a square matrix.

Direct Method

This method is quite easy to visualize, since it has an obvious relationship with the definition of the inverse.

Example

Find the inverse of

$$A = \begin{bmatrix} 4 & 2 \\ 1 & 3 \end{bmatrix}$$

We know that the inverse of A is another matrix A^{-1} such that $AA^{-1} = I$. Let

$$A^{-1} = \begin{bmatrix} b_{11} & b_{12} \\ b_{21} & b_{22} \end{bmatrix}$$

Then, by definition,

$$AA^{-1} = \begin{bmatrix} 4 & 2 \\ 1 & 3 \end{bmatrix} \begin{bmatrix} b_{11} & b_{12} \\ b_{21} & b_{22} \end{bmatrix} = \begin{bmatrix} 1 & 0 \\ 0 & 1 \end{bmatrix}$$

From the definition of equality of matrices, we get

$$4b_{11} + 2b_{21} = 1 \tag{7}$$
$$4b_{12} + 2b_{22} = 0 \tag{8}$$
$$1b_{11} + 3b_{21} = 0 \tag{9}$$
$$1b_{12} + 3b_{22} = 1 \tag{10}$$

From Equation (8), $b_{12} = -\tfrac{1}{2}b_{22}$; when substituted in (10), this gives

$$-\tfrac{1}{2}b_{22} + 3b_{22} = 1$$

or

$$b_{22} = \tfrac{2}{5}$$

$$\therefore \quad b_{12} = -\tfrac{1}{2}b_{22} = -\tfrac{1}{2}(\tfrac{2}{5}) = -\tfrac{1}{5}$$

From Equation (9), $b_{11} = -3b_{21}$; when substituted in (7), this gives

$$4(-3b_{21}) + 2b_{21} = 1$$

or

$$b_{21} = -\tfrac{1}{10}$$

$$\therefore \quad b_{11} = -3b_{21} = -3(-\tfrac{1}{10}) = \tfrac{3}{10}$$

Hence,

$$A^{-1} = \begin{bmatrix} b_{11} & b_{12} \\ b_{21} & b_{22} \end{bmatrix} = \begin{bmatrix} \tfrac{3}{10} & -\tfrac{1}{5} \\ -\tfrac{1}{10} & \tfrac{2}{5} \end{bmatrix}$$

Check:

$$AA^{-1} = \begin{bmatrix} 4 & 2 \\ 1 & 3 \end{bmatrix} \begin{bmatrix} \frac{3}{10} & -\frac{1}{5} \\ -\frac{1}{10} & \frac{2}{5} \end{bmatrix} = \begin{bmatrix} 1 & 0 \\ 0 & 1 \end{bmatrix}$$

It is obvious that finding the inverse of a matrix of order higher than 2×2 would be a rather cumbersome task using the direct method.

Inverting a Matrix by Utilizing Its Determinant and Its Adjoint Matrix

Without giving a formal proof, we define the inverse of a matrix A as

$$A^{-1} = \frac{A_{\text{adj}}}{|A|}$$

where the determinant $|A|$ and the adjoint matrix A_{adj} have been defined earlier.

Example

Find the inverse of

$$A = \begin{bmatrix} 4 & 2 \\ 1 & 3 \end{bmatrix}$$

The determinant of the matrix A is

$$|A| = 12 - 2 = 10$$

and the adjoint of matrix A is

$$A_{\text{adj}} = \begin{bmatrix} C_{11} & C_{21} \\ C_{12} & C_{22} \end{bmatrix} = \begin{bmatrix} 3 & -2 \\ -1 & 4 \end{bmatrix}$$

Hence,

$$A^{-1} = \frac{A_{\text{adj}}}{|A|} = \frac{1}{10} \begin{bmatrix} 3 & -2 \\ -1 & 4 \end{bmatrix} = \begin{bmatrix} \frac{3}{10} & -\frac{1}{5} \\ -\frac{1}{10} & \frac{2}{5} \end{bmatrix}$$

Note that this result is the same as that obtained by the direct method.

Example

Find the inverse of

$$A = \begin{bmatrix} 4 & 0 & 0 \\ 0 & 6 & 2 \\ 2 & 0 & 1 \end{bmatrix}$$

Expanding along the first row,

$$|A| = a_{11}C_{11} + a_{12}C_{12} + a_{13}C_{13}$$
$$= 4(6) + 0 + 0 = 24$$

The adjoint matrix is

$$A_{adj} = \begin{bmatrix} C_{11} & C_{21} & C_{31} \\ C_{12} & C_{22} & C_{32} \\ C_{13} & C_{23} & C_{33} \end{bmatrix} = \begin{bmatrix} 6 & 0 & 0 \\ 4 & 4 & -8 \\ -12 & 0 & 24 \end{bmatrix}$$

Now,

$$A^{-1} = \frac{A_{adj}}{|A|} = \tfrac{1}{24} \begin{bmatrix} 6 & 0 & 0 \\ 4 & 4 & -8 \\ -12 & 0 & 24 \end{bmatrix}$$

$$= \begin{bmatrix} \frac{1}{4} & 0 & 0 \\ \frac{1}{6} & \frac{1}{6} & -\frac{1}{3} \\ -\frac{1}{2} & 0 & 1 \end{bmatrix}$$

4.12   SYSTEMS OF LINEAR EQUATIONS

A system of linear equations may be classified into any one of three categories:

Category 1

The system of linear equations contains n equations in n unknowns.   In this category, we can have one of three cases:

a. The system has a unique solution (see Appendix III).
b. The system is inconsistent and has no solution (see Appendix IV).
c. The system has an infinite number of solutions (see Appendix V).

Category 2

The system of linear equations has more equations than unknowns. In this category, it is possible that no values of the unknown variables may simultaneously satisfy all the equations. Therefore, we usually use some criterion to choose a most desirable set of values. Fitting a line of least squares, for example, is an illustration of solving problems in this category.

Category 3

The system of linear equations has more unknowns than equations. In this case, the system can be solved by assigning an arbitrary value(s) to the excess of unknowns over equations. Thus, we can find an infinite number of solutions to a given system of linear equations in this category. Of these, one or more are chosen in order to obtain an optimal value of some objective function.

The reader will realize that linear-programming problems fall into category 3.

4.13 TERMINOLOGY OF LINEAR-PROGRAMMING SOLUTIONS

With the material covered in Chapters 1 to 3 as background, a general statement of a typical linear-programming problem can be made as follows:

Maximize

$$F(X)^* = c_1x_1 + c_2x_2 + \cdots + c_nx_n$$

subject to the linear structural constraints†

$$a_{11}x_1 + a_{12}x_2 + \cdots + a_{1n}x_n \leq b_1$$
$$a_{21}x_1 + a_{22}x_2 + \cdots + a_{2n}x_n \leq b_2$$
$$\cdots \cdots \cdots \cdots \cdots \cdots \cdots \cdots$$
$$a_{m1}x_1 + a_{m2}x_2 + \cdots + a_{mn}x_n \leq b_m .$$

* This corresponds to the objective function involving profit contribution (see Section 3.3).

† This corresponds to the inequalities containing the technical specifications of Table 3.2.

and subject to the nonnegativity constraints*

$$x_1 \geq 0 \qquad x_2 \geq 0 \qquad \cdots \qquad x_n \geq 0$$

As we observed in Chapter 3, the linear structural constraints of the "less than or equal to" type are transformed into equations by the addition of nonnegative slack variables. When the above problem is stated in the form of equations, we have $n + m$ unknown variables (n structural variables and m slack variables). Thus, when stated in the form of equations, a linear-programming problem is such that the number of equations is less than the number of unknowns. Hence, an infinite number of solutions can be found. Of these, the solution that optimizes the objective function is chosen.

Obtaining a solution to a linear-programming problem involves assigning specific values to the unknown structural variables ($x_1, x_2, \ldots, x_n$) and the unknown slack variables ($x_{n+1}, x_{n+2}, \ldots, x_{n+m}$) without violating the given structural constraints and the nonnegativity constraints. The following terminology, depending upon the number of positive variables involved in a particular solution, is used to identify different types of solutions:

1. *Feasible solution:* any solution containing more than m positive variables and satisfying the linear structural constraints and the nonnegativity constraints†

2. *Basic feasible solution:* any solution containing exactly m positive variables and satisfying the linear structural constraints and the nonnegativity constraints

3. *Degenerate basic feasible solution:* any solution containing less than m positive variables and satisfying the linear structural constraints and the nonnegativity constraints

4. *Optimum solution:* any basic feasible or degenerate basic feasible solution that either maximizes or minimizes the objective function

In the following chapters, while solving linear-programming problems, we shall refer to different solutions in terms of the system of classification given above. We avoided using this terminology in Chapters 2 and 3 because it was felt that some familiarity with matrices and vectors was necessary in order to make the classification more meaningful. The reader should now be able to apply this system of classification to the different solution stages of the graphical and systematic trial-and-error methods.

---

* These are the usual nonnegativity constraints of a linear-programming problem.
† Where m equals the number of linear structural constraints.

The Vector Method

chapter

5

5.1 INTRODUCTION

The reader will recall (Section 4.7) that a set of linear equations can be represented as vectors, and vice versa.   Thus, in so far as a typical linear-programming problem can be stated in terms of linear equations, we can represent the problem by employing vector notation.   The solution of the linear-programming problem, then, can be obtained by performing certain vector operations.   This approach to solving linear-programming problems will be referred to as the *vector method*.

The linear-programming problem of Table 2.1 has now been solved by the graphical method (Chapter 2) and the systematic trial-and-error method (Chapter 3).   In this chapter, the same problem will be solved by the vector method.   A knowledge of the vector method will help the reader to understand the mechanics and rationale of the simplex method (Chapter 6).

For quick reference, the data of Table 2.1 are reproduced in Table 5.1.

Table 5.1 *Process Time by Size and Department*

Department	Size			Capacity per time period
	A	B	C	
Cutting...........................	10.7	5.0	2.0	2,705
Folding...........................	5.4	10.0	4.0	2,210
Packaging........................	0.7	1.0	2.0	445
Profit contribution per unit...........	$10	$15	$20	

95

5.2   VECTOR REPRESENTATION OF THE PROBLEM

Using the arguments of the systematic trial-and-error method, we express the data of Table 5.1 in the form of equations:

$$10.7X + 5Y + 2Z + 1S_1 + 0S_2 + 0S_3 = 2{,}705 \tag{1}$$

$$5.4X + 10Y + 4Z + 0S_1 + 1S_2 + 0S_3 = 2{,}210 \tag{2}$$

$$0.7X + 1Y + 2Z + 0S_1 + 0S_2 + 1S_3 = 445 \tag{3}$$

The objective function* is

$$10X + 15Y + 20Z + 0(S_1 + S_2 + S_3) \tag{4}$$

Writing Equations (1) to (3) in the vector form, we obtain

$$X \begin{bmatrix} 10.7 \\ 5.4 \\ 0.7 \end{bmatrix} + Y \begin{bmatrix} 5 \\ 10 \\ 1 \end{bmatrix} + Z \begin{bmatrix} 2 \\ 4 \\ 2 \end{bmatrix} + S_1 \begin{bmatrix} 1 \\ 0 \\ 0 \end{bmatrix} + S_2 \begin{bmatrix} 0 \\ 1 \\ 0 \end{bmatrix} + S_3 \begin{bmatrix} 0 \\ 0 \\ 1 \end{bmatrix} = \begin{bmatrix} 2{,}705 \\ 2{,}210 \\ 445 \end{bmatrix}$$

If we let

$$P_1 = \begin{bmatrix} 10.7 \\ 5.4 \\ 0.7 \end{bmatrix} \quad P_2 = \begin{bmatrix} 5 \\ 10 \\ 1 \end{bmatrix} \quad P_3 = \begin{bmatrix} 2 \\ 4 \\ 2 \end{bmatrix} \quad P_4 = \begin{bmatrix} 1 \\ 0 \\ 0 \end{bmatrix}$$

$$P_5 = \begin{bmatrix} 0 \\ 1 \\ 0 \end{bmatrix} \quad P_6 = \begin{bmatrix} 0 \\ 0 \\ 1 \end{bmatrix} \quad P_0 = \begin{bmatrix} 2{,}705 \\ 2{,}210 \\ 445 \end{bmatrix}$$

then

$$XP_1 + YP_2 + ZP_3 + S_1P_4 + S_2P_5 + S_3P_6 = P_0 \tag{5}$$

Here, P_1, P_2, and P_3 are the so-called *structural vectors*; P_4, P_5, and P_6 are the *unit vectors*; P_0 is the *constant* or *requirement vector*. Equation (5) states the problem in simple terms. P_0 is a three-component vector which must be expressed as a linear combination of scalar multiples of P_1, P_2, P_3, P_4, P_5, and P_6. In other words, the scalars X, Y, Z, S_1, S_2, and S_3 are to be given nonnegative values such that Equation (5) is satisfied.

---

* The problem, of course, is to maximize the objective function subject to linear constraints (1) to (3) and the nonnegativity constraints $X \geq 0$, $Y \geq 0$, $Z \geq 0$, $S_1 \geq 0$, $S_2 \geq 0$, and $S_3 \geq 0$.

As can be ascertained quickly, and as we noted while solving the same problem by the graphical method, an infinite number of values of the scalars can be found to satisfy Equation (5). Our objective is to choose that set of values which maximizes the profit-contribution function given by (4).

One further point must be emphasized. Since P_0 is a three-dimensional vector, we do not need more than three linearly independent vectors to represent it in a unique fashion. Accordingly, the problem can be solved as follows:

1. Express P_0 in terms of all possible linear combinations of P_1, P_2, P_3, P_4, P_5, and P_6 taken three at a time.
2. Calculate the profit contribution resulting from each such combination, and choose that combination which yields the highest profit.

Theoretically, then, we shall have to test as many as 20 combinations in this simple problem.* This, however, would involve rather lengthy calculations. Instead, to save time and effort and to evolve some systematic method of search, we should like to be able to choose a particular combination of three linearly independent vectors as a starting point and then progressively improve our solution. The method of improvement should be such that combinations giving higher profits than the current program could be immediately identified. In the vector method, this is accomplished in the following manner:

First, an initial program is designed in such a manner that it represents a basic feasible solution.† Second, in order to determine whether the initial or current program can be improved, the net effect on the objective function of introducing one of the nonbasis vectors to replace at least one of the basis vectors is tested.‡ If the objective function can be improved, this replacement is made. In other words, the old basis is replaced by a

---

* By definition, the combination of n things taken r at a time is given by

$$\binom{n}{r} = \frac{n!}{r!(n-r)!}$$

In this case, we have $6!/3!3! = 20$.

† This means, as the reader will recall from Section 4.13, that there are exactly three positive variables (since this is a three-dimensional problem) in the solution. In vector terminology, this means that a linear combination of three vectors is used to represent the requirement vector P_0. Further, the three vectors in the linear combination are the basis vectors, while the remaining three are the nonbasis vectors.

‡ This test is made for *each* of the nonbasis vectors. Then that nonbasis vector which shows the highest improvement potential is introduced in the new basis.

new basis. This testing and replacement process is continued until the optimum set of basis vectors expressing P_0 is determined.

Proceeding in the above fashion (rather than checking each possible combination of the given vectors) reduces the computational work considerably. This, then, is the essence of the vector method.

5.3 ILLUSTRATION OF THE VECTOR METHOD

Designing an Initial Program

As a first step, let us express the requirement vector P_0 as a linear combination of the set of vectors P_4, P_5, and P_6.* Our choice of vectors P_4, P_5, and P_6 as the basis vectors means that we are letting the scalars X, Y, and Z equal zero. In terms of the graphical method, this is equivalent to starting the initial solution at the origin of the three-dimensional space. Physically speaking, this corresponds to producing nothing and thereby letting all the resource capacities stay idle.

In Equation (5), if we let X, Y, and Z equal zero, we obtain

$$0P_1 + 0P_2 + 0P_3 + S_1P_4 + S_2P_5 + S_3P_6 = P_0 \tag{6}$$

or

$$S_1 \begin{bmatrix} 1 \\ 0 \\ 0 \end{bmatrix} + S_2 \begin{bmatrix} 0 \\ 1 \\ 0 \end{bmatrix} + S_3 \begin{bmatrix} 0 \\ 0 \\ 1 \end{bmatrix} = \begin{bmatrix} 2{,}705 \\ 2{,}210 \\ 445 \end{bmatrix}$$

Obviously, if we let $S_1 = 2{,}705$, $S_2 = 2{,}210$, and $S_3 = 445$, the above equation is satisfied. Our initial solution, therefore, is $X = 0$, $Y = 0$,

---

* As we observed in Chapter 4,

$$P_4 = \begin{bmatrix} 1 \\ 0 \\ 0 \end{bmatrix} \qquad P_5 = \begin{bmatrix} 0 \\ 1 \\ 0 \end{bmatrix} \qquad P_6 = \begin{bmatrix} 0 \\ 0 \\ 1 \end{bmatrix}$$

are linearly independent and, thus, comprise a set of *basis vectors* for a three-dimensional space. Hence, the initial basis B is an identity matrix:

$$B = [P_4 \quad P_5 \quad P_6] = \begin{bmatrix} 1 & 0 & 0 \\ 0 & 1 & 0 \\ 0 & 0 & 1 \end{bmatrix}$$

$Z = 0$, $S_1 = 2,705$, $S_2 = 2,210$, and $S_3 = 445$. This is our *first* or *initial* program. Substitution of these values for X, Y, Z, S_1, S_2, and S_3 into the objective function (4) shows that the profit contribution of this program is zero. Equation (6) becomes

$$0P_1 + 0P_2 + 0P_3 + 2,705P_4 + 2,210P_5 + 445P_6 = P_0 \tag{7}$$

Revising the Initial Program

Can we improve our initial program? Since its profit contribution was found to be zero, the answer is obviously in the affirmative. In order to revise the initial program, we must first test the net effect on the objective function of introducing one of the nonbasis vectors P_1, P_2, or P_3 in place of one of the basis vectors P_4, P_5, or P_6. If any such replacement shows improvement potential, then a new set of basis vectors must be determined to form a linear combination for expressing P_0.* This replacement process, it should be emphasized again, must be conducted such that only *one* nonbasis vector is introduced at a time. Thus, in this problem, the basis (for a basic feasible solution) will always consist of three vectors.

Obviously, any one of the nonbasis vectors P_1, P_2, and P_3, if "brought into" the basis, will improve the profit contribution at this stage of the solution. One unit of P_3, for example, if brought in at this stage of the solution, will improve the objective function by $20. The introduction of 1 unit of P_3 (having a profit contribution of $20 per unit) means, as we shall show below, that 2 units of P_4, 4 units of P_5, and 2 units of P_6 must be replaced (P_4, P_5, and P_6 have profit contributions of zero per unit).† Similarly, it can be shown that bringing in 1 unit of P_1 or 1 unit of P_2 will result in a net advantage of $10 or $15, respectively, at this stage of the solution. The point to emphasize is that the *net* advantages of these exchanges are different from stage to stage and thus must be examined separately at each stage of the solution.

Since the per-unit contribution associated with P_3 is greater than that of P_1 or P_2 at this stage, we now propose to replace one of the basis vectors P_4, P_5, or P_6 with P_3. The procedure is explained below.

Since P_3 is itself a three-dimensional vector, it can be expressed as a linear combination of the current basis vectors P_4, P_5, and P_6 in the

---

* The fact that P_0 is expressed as a linear combination of the set of basis vectors means that we are satisfying the linear structural constraints.

† Examine the correspondence between this statement and Equations (14) to (16) of Chapter 3.

initial solution.   To determine the particular linear combination, let

$$aP_4 + bP_5 + cP_6 = P_3$$

where a, b, and c are scalars, or

$$a \begin{bmatrix} 1 \\ 0 \\ 0 \end{bmatrix} + b \begin{bmatrix} 0 \\ 1 \\ 0 \end{bmatrix} + c \begin{bmatrix} 0 \\ 0 \\ 1 \end{bmatrix} = \begin{bmatrix} 2 \\ 4 \\ 2 \end{bmatrix}$$

The above equation is obviously satisfied if we let $a = 2$, $b = 4$, and $c = 2$. Thus, $2P_4 + 4P_5 + 2P_6 = P_3$, which is to say that to introduce 1 unit of P_3 we must remove from the solution 2 units of P_4, 4 units of P_5, and 2 units of P_6.   Further, as can be seen from the profit function, such an exchange will increase profit by $+1(20) - 2(0) - 4(0) - 2(0) = \$20$ for every unit of P_3 brought into the solution.   This exchange being advantageous, we shall carry it to the limit.   In other words, we shall keep on bringing units of P_3 into the solution until *one* of the vectors P_4, P_5, or P_6 is removed from the solution.   Let us say that at most h units of P_3 can be brought in.   Then

$$2hP_4 + 4hP_5 + 2hP_6 = hP_3$$

or

$$2hP_4 + 4hP_5 + 2hP_6 - hP_3 = 0 \tag{8}$$

Since the subtraction of zero from a given equation does not change anything, let us subtract (8) from (7).   Then

$$0P_1 + 0P_2 + hP_3 + (2{,}705 - 2h)P_4 + (2{,}210 - 4h)P_5$$
$$+ (445 - 2h)P_6 = P_0 \tag{9}$$

In Equation (9), P_0 has been expressed as a linear combination of P_3, P_4, P_5, and P_6.   However, as stated previously, we need only three vectors in the solution.   One of the old basis vectors P_4, P_5, or P_6 should therefore be reduced to zero.   To reduce P_4 to zero, let

$$2{,}705 - 2h = 0 \qquad \text{or} \qquad h = 1{,}352.5$$

To reduce P_5 to zero, let

$$2,210 - 4h = 0 \qquad \text{or} \qquad h = 552.5$$

To reduce P_6 to zero, let

$$445 - 2h = 0 \qquad \text{or} \qquad h = 222.5 \checkmark$$

This means that the maximum number of units of P_3 that can be brought into the solution is 222.5.   In other words, $h = 222.5$ is the controlling or limiting number and maximum allowable value that can be inserted into Equation (9) without making any term on the left-hand side negative.*
Letting $h = 222.5$ in Equation (9), we have

$$0P_1 + 0P_2 + 222.5P_3 + 2,260P_4 + 1,320P_5 + 0P_6 = P_0 \qquad (10)$$

The *second* program, therefore, is

$$X = 0 \qquad Y = 0 \qquad Z = 222.5 \qquad S_1 = 2,260 \qquad S_2 = 1,320 \qquad S_3 = 0$$

This program yields a profit of $20(222.5) = \$4,450.00$.
Now the basis vectors are P_3, P_4, and P_5, and the nonbasis vectors are P_1, P_2, and P_6.

Revision of the Second Program

Is the second program our optimal program?   As stated previously, we can answer this question only by testing the net effect on the objective function of bringing in one of the nonbasis vectors P_1, P_2, or P_6 to replace one of the present basis vectors P_3, P_4, or P_5.   Representing P_2 as a linear combination of scalar multiples of the three basis vectors (P_3, P_4, and P_5) now in the solution, we obtain

$$aP_3 + bP_4 + cP_5 = P_2$$

where a, b, and c are scalars, or

$$a \begin{bmatrix} 2 \\ 4 \\ 2 \end{bmatrix} + b \begin{bmatrix} 1 \\ 0 \\ 0 \end{bmatrix} + c \begin{bmatrix} 0 \\ 1 \\ 0 \end{bmatrix} = \begin{bmatrix} 5 \\ 10 \\ 1 \end{bmatrix}$$

---

* We cannot let any term in Equation (9) become negative, as this would violate the nonnegativity constraints.

The above vector equation can be translated into the following equations:

$$2a + 1b + 0c = 5$$
$$4a + 0b + 1c = 10$$
$$2a + 0b + 0c = 1$$

Solving this system of equations, we get

$$a = \tfrac{1}{2} \qquad b = 4 \qquad c = 8$$

or

$$\tfrac{1}{2}P_3 + 4P_4 + 8P_5 = P_2$$

In other words, to bring in 1 unit of P_2 we shall have to remove from the solution $\tfrac{1}{2}$ unit of P_3, 4 units of P_4, and 8 units of P_5. This exchange will have the following effect on the profit function:

$$+1(15) - \tfrac{1}{2}(20) - 4(0) - 8(0) = +5 \text{ dollars}$$

In so far as this is a profitable exchange, we shall replace at least one of the vectors now in the solution, namely, P_3, P_4, or P_5, with P_2.*

Following the same argument as before, let us say that at most k units of P_2 can be brought in without violating the nonnegativity constraints. Then

$$\tfrac{1}{2}kP_3 + 4kP_4 + 8kP_5 = kP_2$$

or

$$\tfrac{1}{2}kP_3 + 4kP_4 + 8kP_5 - kP_2 = 0 \tag{11}$$

---

* Actually, the relative exchange profitabilities of all the nonbasis vectors (here P_1, P_2, and P_6) should have been compared, and the nonbasis vector with the highest net advantage chosen to make the replacement. But here, as soon as it was found that the nonbasis vector P_2 had improvement potential, we decided to obtain a new basis to include P_2. A complete set of comparisons among all the nonbasis vectors was not made because our intent is only to focus on the vector method as a forerunner of the simplex method, which does make *all* such comparisons at each stage of the solution. As it turns out, P_2 is the correct choice.

We subtract (11) from (10):

$$0P_1 + kP_2 + (222.5 - \tfrac{1}{2}k)P_3 + (2{,}260 - 4k)P_4$$
$$+ (1{,}320 - 8k)P_5 + 0P_6 = P_0 \quad (12)$$

In Equation (12), P_0 has been expressed as a linear combination of P_2, P_3, P_4, and P_5. However, we really need only three vectors in the solution, and one of the vectors P_3, P_4, or P_5 should be reduced to zero. To reduce P_3 to zero, let

$$222.5 - \tfrac{1}{2}k = 0 \quad \text{or} \quad k = 445$$

To reduce P_4 to zero, let

$$2{,}260 - 4k = 0 \quad \text{or} \quad k = 565$$

To reduce P_5 to zero, let

$$1{,}320 - 8k = 0 \quad \text{or} \quad k = 165 \;\checkmark$$

Thus, $k = 165$ is the limiting case. That is, we cannot bring in more than 165 units of P_2 without violating the nonnegativity constraints. Substituting $k = 165$ in Equation (12), we get

$$0P_1 + 165P_2 + 140P_3 + 1{,}600P_4 + 0P_5 + 0P_6 = P_0 \quad\quad (13)$$

The *third* program, therefore, is

$$X = 0 \quad\quad Y = 165 \quad\quad Z = 140 \quad\quad S_1 = 1{,}600 \quad\quad S_2 = 0 \quad\quad S_3 = 0$$

This program yields a profit of $15(165) + 20(140) = \$5{,}275$.

The basis vectors of this program are P_2, P_3, and P_4; the nonbasis vectors are P_1, P_5, and P_6.

Revision of the Third Program

Again we pose the same question: Can we further improve the current program? Note that any improvement in the present program can be made only by replacing one of the basis vectors now in the solution (P_2,

P_3, P_4) with one of the nonbasis vectors *not* in the solution (P_1, P_5, P_6). Let us determine the consequences of bringing in P_1.

Expressing P_1 as a linear combination of the basis vectors (P_2, P_3, and P_4), we have

$$aP_2 + bP_3 + cP_4 = P_1$$

where a, b, and c are scalars; then

$$a\begin{bmatrix} 5 \\ 10 \\ 1 \end{bmatrix} + b\begin{bmatrix} 2 \\ 4 \\ 2 \end{bmatrix} + c\begin{bmatrix} 1 \\ 0 \\ 0 \end{bmatrix} = \begin{bmatrix} 10.7 \\ 5.4 \\ 0.7 \end{bmatrix}$$

The above vector equation is translated into the following equations:

$$5a + 2b + c = 10.7$$
$$10a + 4b + 0c = 5.4$$
$$1a + 2b + 0c = 0.7$$

Solving this system of equations, we get

$$a = \tfrac{1}{2} \qquad b = \tfrac{1}{10} \qquad c = 8$$

Hence,

$$\tfrac{1}{2}P_2 + \tfrac{1}{10}P_3 + 8P_4 = P_1$$

In other words, to bring in 1 unit of P_1, we must remove $\tfrac{1}{2}$ unit of P_2, $\tfrac{1}{10}$ unit of P_3, and 8 units of P_4 from the solution. For the effect of this exchange on profit, we have

$$+10 - \tfrac{1}{2}(15) - \tfrac{1}{10}(20) - 8(0) = +\tfrac{1}{2} \text{ dollar}$$

This indicates that the exchange is profitable and, therefore, should be pressed to its limit. Proceeding as before, we assume that the maximum number of units of P_1 that can be brought in is m units. Then

$$\tfrac{1}{2}mP_2 + \tfrac{1}{10}mP_3 + 8mP_4 = mP_1$$

or

$$\tfrac{1}{2}mP_2 + \tfrac{1}{10}mP_3 + 8mP_4 - mP_1 = 0 \qquad (14)$$

We subtract (14) from (13):

$$mP_1 + (165 - \tfrac{1}{2}m)P_2 + (140 - \tfrac{1}{10}m)P_3$$
$$+ (1{,}600 - 8m)P_4 + 0P_5 + 0P_6 = P_0 \quad (15)$$

As explained previously, we know that at least one of the vectors P_2, P_3, or P_4 in (15) must be reduced to zero. To reduce P_2 to zero, let

$$165 - \tfrac{1}{2}m = 0 \qquad \text{or} \qquad m = 330$$

To reduce P_3 to zero, let

$$140 - \tfrac{1}{10}m = 0 \qquad \text{or} \qquad m = 1{,}400$$

To reduce P_4 to zero, let

$$1{,}600 - 8m = 0 \qquad \text{or} \qquad m = 200 \checkmark$$

We note that $m = 200$ is the limiting case. Substituting $m = 200$ in Equation (15), we get

$$200P_1 + 65P_2 + 120P_3 + 0P_4 + 0P_5 + 0P_6 = P_0$$

The *fourth* program, therefore, is

$$X = 200 \qquad Y = 65 \qquad Z = 120 \qquad S_1 = 0 \qquad S_2 = 0 \qquad S_3 = 0$$

This program yields a profit contribution of

$$10(200) + 15(65) + 20(120) = \$5{,}375$$

Having solved the same problem earlier, we know that this is the optimum solution. However, we can test the effect on the objective function of the replacement of any one of the present basis vectors (P_1, P_2, P_3) by any one of the nonbasis vectors (P_4, P_5, P_6).   The tests are summarized below.

Effect of Bringing in P_4

Let $P_4 = aP_1 + bP_2 + cP_3$; then

$$\begin{bmatrix} 1 \\ 0 \\ 0 \end{bmatrix} = a \begin{bmatrix} 10.7 \\ 5.4 \\ 0.7 \end{bmatrix} + b \begin{bmatrix} 5 \\ 10 \\ 1 \end{bmatrix} + c \begin{bmatrix} 2 \\ 4 \\ 2 \end{bmatrix}$$

whence

$$a = \tfrac{1}{8} \qquad b = -\tfrac{1}{16} \qquad c = -\tfrac{1}{80}$$

The effect on profit is

$$+0 - \tfrac{1}{8}(10) + \tfrac{1}{16}(15) + \tfrac{1}{80}(20) = -\tfrac{5}{80} = -\tfrac{1}{16}$$

In other words, for every unit of P_4 brought into the solution at this stage, total profit would be reduced by $\tfrac{1}{16}$ dollar.

Effect of Bringing in P_5

Let $P_5 = aP_1 + bP_2 + cP_3$; then

$$\begin{bmatrix} 0 \\ 1 \\ 0 \end{bmatrix} = a \begin{bmatrix} 10.7 \\ 5.4 \\ 0.7 \end{bmatrix} + b \begin{bmatrix} 5 \\ 10 \\ 1 \end{bmatrix} + c \begin{bmatrix} 2 \\ 4 \\ 2 \end{bmatrix}$$

whence

$$a = -\tfrac{1}{16} \qquad b = \tfrac{5}{32} \qquad c = -\tfrac{9}{160}$$

The effect on profit is

$$+0 + \tfrac{1}{16}(10) - \tfrac{5}{32}(15) + \tfrac{9}{160}(20) = -\tfrac{95}{160}$$

In other words, for every unit of P_5 brought into the solution at this stage, total profit would be reduced by $\tfrac{19}{32}$ dollar.

Effect of Bringing in P_6

Let $P_6 = aP_1 + bP_2 + cP_3$; then

$$\begin{bmatrix} 0 \\ 0 \\ 1 \end{bmatrix} = a \begin{bmatrix} 10.7 \\ 5.4 \\ 0.6 \end{bmatrix} + b \begin{bmatrix} 5 \\ 10 \\ 1 \end{bmatrix} + c \begin{bmatrix} 2 \\ 4 \\ 2 \end{bmatrix}$$

whence

$$a = 0 \qquad b = -\tfrac{1}{4} \qquad c = \tfrac{5}{8}$$

The effect on profit is

$$+0 + 0(10) + \tfrac{1}{4}(15) - \tfrac{5}{8}(20) = -\tfrac{35}{4}$$

In other words, for every unit of P_6 to be brought into the solution at this stage, total profit would be reduced by $\tfrac{35}{4}$ dollars.

Table 5.2

Program	Basis vectors	Nonbasis vectors	Profit contribution
1	P_4, P_5, P_6	P_1, P_2, P_3	0
2	P_4, P_5, P_3	P_1, P_2, P_6	\$4,450
3	P_4, P_2, P_3	P_1, P_5, P_6	\$5,275
4	P_1, P_2, P_3	P_4, P_5, P_6	\$5,375

We therefore note that any change in the fourth program will decrease rather than increase the value of the profit function. Hence, the fourth program is the optimum program.

Given in Table 5.2 is a summary of the various solution stages of the problem as solved by the vector method. The reader should compare this summary with the one given in Table 3.3.

5.4 PROCEDURE SUMMARY FOR THE VECTOR METHOD (MAXIMIZATION CASE)

Step 1   Formulate the Problem

a. Translate the technical specifications of the problem into inequalities, and make a precise statement of the objective function.

b. Convert the inequalities into equalities by the addition of nonnegative slack variables. Attach a per-unit profit of zero to each of these slack variables or "imaginary" products.
c. Write the equations obtained in (*b*) in the form of vectors.

Step 2   Design a Program (*Choose a Set of Basis Vectors*)

a. Express the requirement vector P_0 as a linear combination of a set of independent vectors (the first program is always obtained by expressing P_0 as a linear combination of unit vectors). This set of vectors forms the basis.
b. Express the program in the form of a vector equation containing all basis and nonbasis vectors with proper scalars.

Step 3   Revise the Program (*Choose a New Set of Basis Vectors*)

a. *Identify the incoming vector.* Express all the nonbasis vectors as linear combinations of the basis vectors, and determine the exchange profitabilities of all the nonbasis vectors. If none of the nonbasis vectors shows any exchange profitability, the problem is solved and no revision is needed. Otherwise the nonbasis vector with the highest exchange profitability is chosen to become one of the new basis vectors.
b. *Determine the maximum units of the incoming vector.* Multiply the exchange equation of the incoming vector by some unknown scalar k and arrange it in such a manner that its right-hand side is zero. Subtract this exchange equation from the program equation of step 2. Then determine the maximum units of the incoming vector (the value of some constant k) so that at least one of the current basis vectors acquires zero as its scalar. This will lead to a new set of basis vectors and hence a new program—expressed in the form of a vector equation.

Step 4   Obtain the Optimum Program

Repeat steps 3 and 4 until an optimal program has been designed. An optimal program has been reached when the exchange profitabilities of all the nonbasis vectors are zero or negative.

5.5   THE VECTOR METHOD (MINIMIZATION CASE)

The procedure for solving a linear-programming problem in which the objective function is to be minimized is exactly the same as given above except that (1) the nonbasis vector whose exchange with the current basis vectors results in the highest cost reduction is chosen as the incoming vector and (2) an optimal solution has been reached when the exchange of each nonbasis vector in the optimal program does not decrease the value of the objective function.

The reader is encouraged to design a minimization problem and solve it by applying the vector method.

6

The Simplex
Method. I

6.1 INTRODUCTION

Of the various methods of solving linear-programming problems, the simplex method is the most general and powerful. Actually, the graphical method (Chapter 2), the systematic trial-and-error method (Chapter 3), and the vector method (Chapter 5) were presented mainly to give the reader a "feel" for linear-programming problems and to acquaint him with some of the technical terminology so essential in understanding the rationale and mechanics of the simplex method. Otherwise, in actual practice, linear-programming problems of any significance are usually solved by application of the simplex method.

*The simplex method is based on the property that the optimum solution to a linear-programming problem, if it exists, can always be found in one of the basic feasible solutions.** Thus, in the simplex method, the first step is always to obtain a basic feasible solution. As explained in the vector method (Chapter 5), this means that we obtain a set of basis vectors and a set of nonbasis vectors. Further, this means that the requirement vector P_0 is expressed as a linear combination of the basis vectors. This, in effect, gives us a solution to the problem. This solution is then tested for optimality by examining the net effect on the linear objective function of introducing one of the nonbasis vectors to replace at least one of the basis vectors. If any improvement potential is noted, the replacement is made, always by introducing only one nonbasis vector at a time. As explained in the last chapter, this replacement results in a *new* basis.

So far we have not indicated anything that is different from the vector

---

* If the problem is degenerate, the optimum solution is a degenerate basic feasible solution.

method. However, as we shall see below, the beauty of the simplex method lies in the fact that the relative exchange profitabilities of all the nonbasis vectors can be determined simultaneously* and easily; the replacement process is such that the new basis does not violate the feasibility of the solution.

The simplex method, as we shall see below, is quite simple and mechanical in nature. The steps of the simplex method are repeated until a finite optimum solution, if it exists, is determined. Otherwise, the method indicates either that the given linear-programming problem has no solution or that no finite maximum exists.

6.2   THE PROBLEM (A MAXIMIZATION CASE)

To fix ideas and to facilitate comparisons with the other methods of solving linear-programming problems, we shall solve the problem of Table 2.1 by the simplex method. For quick reference, the data of Table 2.1 are reproduced as Table 6.1.

Table 6.1   *Process Time by Size and Department*

Department	Size			Capacity per time period
	A	B	C	
Cutting...........................	10.7	5.0	2.0	2,705
Folding...........................	5.4	10.0	4.0	2,210
Packaging.........................	0.7	1.0	2.0	445
Profit contribution per unit............	$10	$15	$20	

As in the other methods, our first step is to translate the technical data into inequalities:

$$10.7X + 5Y + 2Z \le 2{,}705$$

$$5.4X + 10Y + 4Z \le 2{,}210$$

$$0.7X + 1Y + 2Z \le 445$$

By the addition of slack variables S_1, S_2, and S_3, these inequalities can

---

* In the vector method as presented in Chapter 5, the net effect of each nonbasis vector was determined individually.

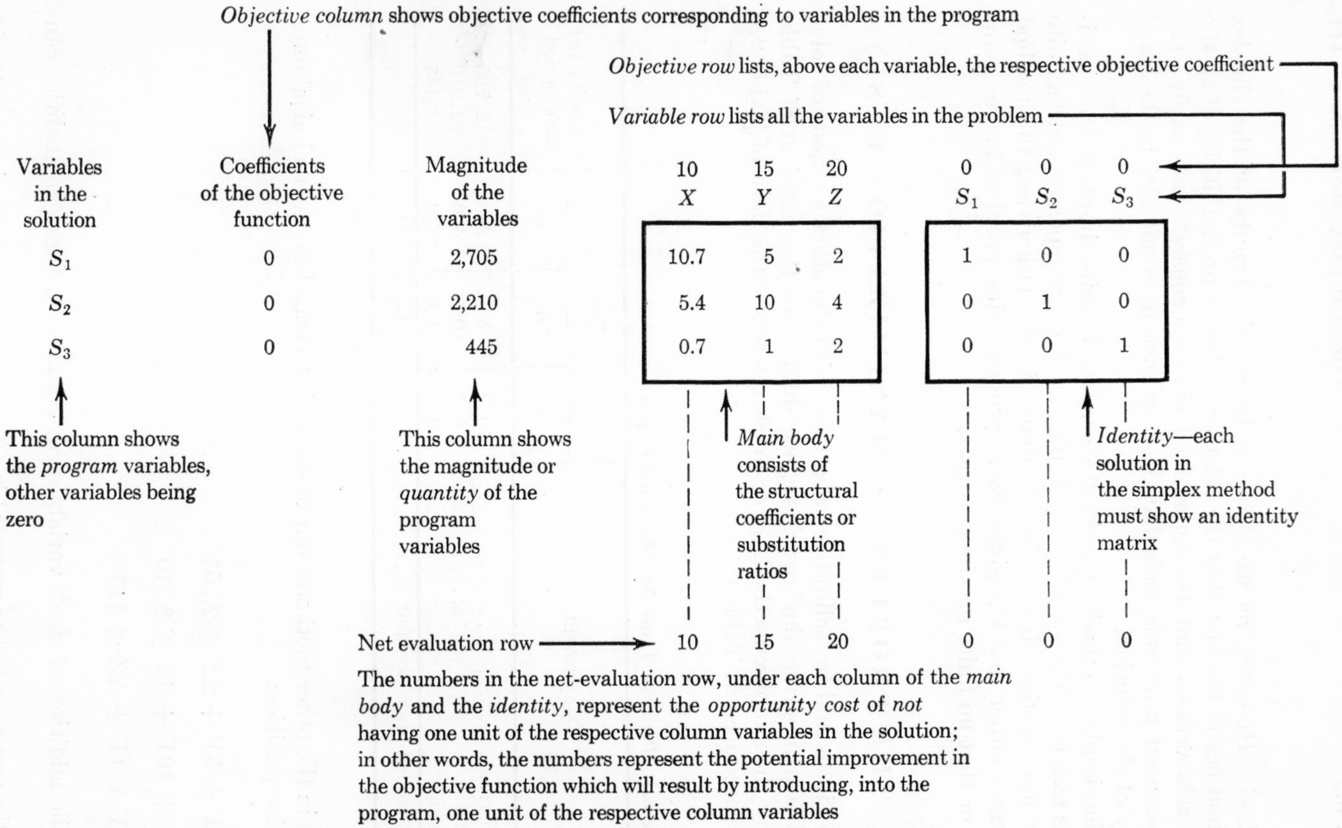

Objective column shows objective coefficients corresponding to variables in the program

Objective row lists, above each variable, the respective objective coefficient

Variable row lists all the variables in the problem

Variables in the solution	Coefficients of the objective function	Magnitude of the variables	10	15	20	0	0	0
			X	Y	Z	S_1	S_2	S_3
S_1	0	2,705	10.7	5	2	1	0	0
S_2	0	2,210	5.4	10	4	0	1	0
S_3	0	445	0.7	1	2	0	0	1
			10	15	20	0	0	0

This column shows the *program* variables, other variables being zero

This column shows the magnitude or *quantity* of the program variables

Main body consists of the structural coefficients or substitution ratios

Identity—each solution in the simplex method must show an identity matrix

Net evaluation row ⟶

The numbers in the net-evaluation row, under each column of the *main body* and the *identity*, represent the *opportunity cost of not* having one unit of the respective column variables in the solution; in other words, the numbers represent the potential improvement in the objective function which will result by introducing, into the program, one unit of the respective column variables

Figure 6.1 Nomenclature of the simplex tableau.

be converted into equations. The slack variables may be given the familiar physical interpretation in which the capacities of cutting, folding, and packaging departments not utilized in producing products X, Y, Z are, respectively, used to produce imaginary products S_1, S_2, S_3, each giving a per-unit profit contribution of zero.

Our problem, then, can be stated as follows:

Maximize

$$10X + 15Y + 20Z + 0S_1 + 0S_2 + 0S_3$$

subject to

$$10.7X + 5Y + 2Z + 1S_1 + 0S_2 + 0S_3 = 2{,}705$$

$$5.4X + 10Y + 4Z + 0S_1 + 1S_2 + 0S_3 = 2{,}210$$

$$0.7X + 1Y + 2Z + 0S_1 + 0S_2 + 1S_3 = 445$$

and $X \geq 0$, $Y \geq 0$, $Z \geq 0$, $S_1 \geq 0$, $S_2 \geq 0$, $S_3 \geq 0$.

The simplex method, which, in effect, is a concentrated and more efficient arrangement of the vector method, proceeds to solve the above problem by designing and redesigning successively better basic feasible solutions until an optimum solution is obtained. Each program, as we shall see below, is given in the form of a matrix or tableau. Although there are various forms for a simplex tableau, we shall follow the one given in Figure 6.1, which contains an initial program for the problem given in Table 6.1.* The nomenclature of the simplex tableau, as identified in Figure 6.1, will be followed throughout this book. If the rows and columns of a given tableau are labeled in a different fashion, it is solely for the sake of convenience. To interpret the basic role and significance of these, we can always refer back to Figure 6.1.

6.3   DESIGNING THE INITIAL PROGRAM

As in the systematic trial-and-error method and the vector method, *the first program in the simplex method is that which involves only the slack variables.* This program is summarized in Tableau 6.I.

The interpretation of the data in Tableau 6.I must be fully grasped in order that the simplex method be understood. Let us, therefore, discuss

* See Section 6.3.

prid 6 units

Tableau 6.I

Program	Profit per unit	Quantity	$10 X	$15 Y	$20 Z	$0 S_1	$0 S_2	$0 S_3
S_1	0	2,705	10.7	5	2	1	0	0
S_2	0	2,210	5.4	10	4	0	1	0
S_3	0	445	0.7	1	2	0	0	1

prod to be produced   amt of units that can be produced

the contents of Tableau 6.I.   Other simplex tableaux will have similar interpretations.

1. In the column labeled "Program" are listed the particular variables in the solution (products being produced).*   Thus, in our first program, we are producing only S_1, S_2, and S_3.
2. In the column labeled "Profit per unit" are listed the coefficients (in the objective function) of the variables included in the specific program.   Thus, the coefficients of S_1, S_2, and S_3, which are included in the initial program, are listed in the "Profit per unit" column.   As can be ascertained from the objective function, the coefficients of S_1, S_2, and S_3 are all zero.
3. In the column labeled "Quantity" are listed the magnitudes of the variables included in the solution (quantities of the products being produced in the program).   Since our initial program consists in producing 2,705 units of S_1, 2,210 units of S_2, and 445 units of S_3, these values are listed in the "Quantity" column.
4. The total profit contribution resulting from a given program can be calculated by multiplying corresponding entries in the "Profit per unit" column and the "Quantity" column and adding the products. Thus, in our first program, total profit contribution is zero [(0)2,705 + (0)2,210 + (0)445].
5. Numbers in the main body (entries in columns X, Y, and Z) can be interpreted to mean physical ratios of substitution at a particular stage of the solution process.   In the first tableau, if constructed in the above fashion (i.e., include in the program nothing but the slack variables), these physical ratios of substitution correspond exactly to the given technical specifications.   For example, the number 10.7

* The *basis* variables.

gives the rate of substitution between X and S_1.* In other words, if we wish to produce 1 unit of X, then 10.7 units of S_1 must be "sacrificed." That is, available cutting capacity will be reduced by 10.7 units. The numbers 5.4 and 0.7 have similar interpretations. By the same token, to produce 1 unit of Y, we must "sacrifice" 5 units of S_1, 10 units of S_2, and 1 unit of S_3.

6. Like the numbers in the main body, the entries in the identity (columns S_1, S_2, and S_3) can be interpreted as physical ratios of exchange. Thus, the numbers in column S_1 represent, respectively, the ratios of exchange between S_1 and the variables S_1, S_2, and S_3 in the solution.

7. The numbers at the top of the columns of the main body and identity represent the coefficients of the respective variables in the objective function.

6.4   TESTING THE OPTIMALITY OF THE CURRENT PROGRAM

In so far as the total profit contribution resulting from our initial program is zero, it can obviously be improved and, hence, is not the optimal program. In any case, the signal that an improvement in the current program can be made and that the optimal solution has not been obtained is given by the entries in the so-called *net-evaluation row* or *base row* (see Figure 6.1) of a given tableau. Let us explain.

Assume that we wish to change the program in Tableau 6.I by introducing (producing) 1 unit of X. This would, as explained previously, involve sacrificing 10.7 units of S_1, 5.4 units of S_2, and 0.7 unit of S_3. The effect of this exchange on the profit function would be

$$+1(10) - 10.7(0) - 5.4(0) - 0.7(0) = +10$$

In other words, the introduction of 1 unit of X at this stage of the solution will increase the value of the profit function by \$10. Thus, the opportunity cost of *not* having this unit of X in our solution is \$10. It is this number that is entered in the net-evaluation row under column X. Similarly, as can easily be ascertained, the opportunity costs of *not* having products Y and Z in our solution at this stage are \$15 and \$20 per unit, respectively. This is the significance of the numbers in the net-evaluation row. The mechanics of the calculation of the net-evaluation row of any tableau is given below.

* Note that 10.7 lies at the intersection of column X and row S_1.

> *To get a number in the net-evaluation row under any column, multiply the entries in that column by the corresponding numbers in the objective column, and add the products. Then subtract this sum from the number listed in the objective row at the top of this column.*

The numbers in the net-evaluation row, as the above discussion indicates, represent the potential improvement in the objective function which will result from the introduction into the program of 1 unit of each of the respective column variables. Thus, by definition, these numbers represent the *opportunity costs** of *not* having 1 unit of each of the respective column variables in the solution. Since we are dealing with a linear-programming model which assumes *certainty*, the presence of any

Tableau 6.I

Program	Profit per unit	Quantity	10 X	15 Y	20 Z	0 S_1	0 S_2	0 S_3	
S_1	0	2,705	10.7	5	2	1	0	0	$\frac{2,705}{2} = 1,352.5$
S_2	0	2,210	5.4	10	4	0	1	0	$\frac{2,210}{4} = 552.5$
S_3	0	445	0.7	1	2	0	0	1	$\frac{445}{2} = 222.5 \checkmark$
Net-evaluation row:			10	15	20	0	0	0	

Key number

Key row (outgoing variable)

Key column (incoming variable)

positive opportunity cost in the net-evaluation row of a given tableau indicates that an optimum solution does not exist. *Hence, any positive number in the net-evaluation row is indicative of the presence of positive opportunity cost and implies that a better program can be designed.* This is the criterion that will be used in this book for obtaining an optimum solution of a *maximizing* type of linear-programming problem.†

The net-evaluation row for Tableau 6.I is calculated and the net-evaluation numbers for the variables are listed at the base of Tableau 6.I. An examination of the net-evaluation numbers shows the existence of positive opportunity cost in the initial program. Our first program, therefore, is not an optimum program.

---

* The costs associated with not following the best course of action.
† See Section 7.2 for the criterion of optimality for a minimization problem.

6.5 REVISION OF THE CURRENT PROGRAM

Identification of the Key Column

The three positive numbers (10, 15, 20) in the net-evaluation row of Tableau 6.I indicate, respectively, the magnitudes of the opportunity costs of *not* including 1 unit of variables (products) X, Y, and Z in this program. Since the highest opportunity cost falls under column Z, the variable (product) Z should be brought into the program first. Column Z, then, is called the *key column*.

The column under which falls the largest positive opportunity cost forms the key column.

The reason for calling this column the key column is obvious. It is the variable (product) of this column which is to be brought in the solution, thus providing a "key" in obtaining the revised program.

Identification of the Key Row and the Key Number

After we have decided to "bring in" variable (product) Z to replace at least one of the variables (products) in the current program (S_1, S_2, or S_3), the question becomes: How many units of Z can be brought in without exceeding the existing capacity of any one of the resources? In linear-programming terms, this means that we must calculate the maximum allowable number of units of Z that can be brought into the program without violating the nonnegativity constraints. If we examine Tableau 6.I, we note that to bring in 1 unit of Z, we must "sacrifice" 2 units of S_1, 4 units of S_2, and 2 units of S_3. In so far as we are currently producing only 2,705 units of S_1, it is clear that no more than 1,352.5 units $(2,705/2 = 1,352.5)$ of Z can be brought in without violating the capacity restriction of the cutting department. Similarly, at this stage of the solution, the production of Z is limited to 552.5 units $(2,210/4 = 552.5)$ and 222.5 units $(445/2 = 222.5)$ by the available capacities of the folding department and packaging department, respectively. The limiting case, therefore, arises from row S_3 in Tableau 6.I. This, then, is our *key row*, and 222.5 units is the maximum quantity of the product Z that can be produced, at this stage of the solution, without violating the nonnegativity constraints. The mechanics of the identification of the key row is given below.

Divide the entries under the "Quantity" column by the corresponding nonnegative* entries of the key column, and compare these ratios.   *The row in which the smallest ratio falls is the key row.*   The reason for calling this row the key row is evident.   It is this row that limits the magnitude of the incoming variable (product).   Thus, it provides a "key" in determining the revised program.

The calculations which help identify the key row in Tableau 6.I are

For the cutting department (row S_1):

$$\frac{2,705}{2} = 1,352.5 \text{ units}$$

For the folding department (row S_2):

$$\frac{2,210}{4} = 552.5 \text{ units}$$

For the packaging department (row S_3):

$$\frac{445}{2} = 222.5 \text{ units } \checkmark$$

While going through the simplex algorithm, it is convenient to place such calculations on the extreme right-hand side of a given tableau.   For purposes of identification, we shall refer to the results of these calculations as the *replacement ratios*.   The limiting quantity of the incoming variable (product) is then identified by a simple check mark ($\checkmark$), as has been done in Tableau 6.I.   Obviously, the limiting quantity is given by the *lowest* replacement ratio.

Once the key row and the key column have been determined, the identification of the key number is a simple matter.   *The number which lies at the intersection of the key row and the key column of a given tableau is the key number.*   Thus, in Tableau 6.I, the key number is 2.   The reason for calling this number the key number lies in the fact, as we shall see later, that a mere division of the key row by this number gives us the corresponding row in the next tableau (for the revised program).

---

* A negative entry in the key column, when interpreted as a physical ratio of exchange, would mean that the introduction of the key-column variable increases rather than decreases the magnitude of the row variable in which this negative entry exists.   The current magnitude of this row variable, therefore, would provide no limit to the introduction of the key-column variable.   Hence, in identifying the key row, only the positive ratios of substitution need be examined.

6.6 DERIVATION OF TABLEAU 6.II

The identification of the key column and the key row has shown us that the variable (product) Z will replace variable (product) S_3 and that no more than 222.5 units of Z can be produced under the current capacity restrictions. Our next task is to determine the exact composition of the remainder of the revised program. In other words, we must find the reductions in S_1 and S_2 due to the fact that 222.5 units of Z are to be included in the revised program.

Since it takes 2 units of packaging-department capacity to produce 1 unit of Z, it is evident that all 445 units (222.5 × 2) of the packaging capacity are exhausted. However, as we noted earlier, the production of 1 unit of Z also takes 2 units of cutting-department capacity and 4 units of folding-department capacity. Thus, the remaining capacity of the cutting department is $2{,}705 - (222.5 \times 2) = 2{,}260$, and the remaining capacity of the folding department is $2{,}210 - (222.5 \times 4) = 1{,}320$ units.

Another way to say the same thing is that our second program calls for producing $Z = 222.5$ units, $S_1 = 2{,}260$ units, $S_2 = 1{,}320$ units, $S_3 = 0$, $X = 0$, and $Y = 0$. The three products (S_1, S_2, and Z) included in the second program are listed in the "Program" column in Tableau 6.II.

As the reader will recall, this second program corresponds to using P_3, P_4, and P_5 as the basis vectors in the vector method presented earlier. Further, it should be emphasized that the ratios of substitution among different variables at this stage of the solution are those that were obtained in Equations (17) to (19) of the systematic trial-and-error method (Chapter 3). Since the main body and the identity of a simplex tableau contain these very ratios of substitution, we could use the information of Equations (17) to (19) to build our Tableau 6.II.* The simplex method, however, provides a mechanics with which the task can be simply accomplished. Let us illustrate.

Tableau 6.II, which will contain our second program, is to be derived from Tableau 6.I, which represented our initial program. In so far as the number of variables in the solution from one program to the next remains constant, each simplex tableau formed in the process of solving a linear-programming problem contains a fixed number of rows.† Further, any

---

* Equations (17) to (19) of Chapter 3 are rearranged in the following form:

$$222.5 = 0.35X + 0.5Y + 1Z + 0S_1 + 0S_2 + 0.5S_3$$

$$2{,}260 = 10X + 4Y + 0Z + 1S_1 + 0S_2 - 1S_3$$

$$1{,}320 = 4X + 8Y + 0Z + 0S_1 + 1S_2 - 2S_3$$

† This corresponds to the fact that the number of basis vectors, from one basis to another, remains constant.

given tableau, during the solution stages, has two types of rows: (1) the key row and (2) the nonkey rows. Thus, to derive a new tableau from an old tableau, all we have to do is to establish rules of transformation for these two types of rows. These rules of transformation, presented below, form the mechanical foundations of the simplex method. The rules are to be applied to the entire set of entries of each row, starting with and to the right of the "Quantity" column.

Transformation of the Key Row

The rule for transforming the key row is: *Divide all the numbers in the key row by the key number.* The resulting numbers form the corresponding row (to be placed in exactly the same position) in the next tableau.

Thus, the third row in Tableau 6.II (row Z) is derived from the third row in Tableau 6.I (row S_3) by simply dividing all the numbers in row S_3 by 2 (the key number). The new row Z (Tableau 6.II) is

222.5	0.35	0.5	1	0	0	0.5

Transformation of the Nonkey Rows

The rule for transforming a nonkey row is: *Subtract from the old row number (in each column) the product of the corresponding key-row number and the corresponding fixed ratio formed by dividing the old row number in the key column by the key number.* The result will give the corresponding new row number.

The above rule can be placed in the following equation:

New row number = old row number

$$- \left(\begin{array}{c} \text{corresponding number} \\ \text{in key row} \end{array} \times \begin{array}{c} \text{corresponding} \\ \text{fixed ratio} \end{array} \right)$$

where

$$\text{Fixed ratio} = \frac{\text{old row number in key column}}{\text{key number}}$$

Thus, the new row S_1 for Tableau 6.II is derived as follows (corresponding fixed ratio = 2/2 = 1):

$$\begin{matrix} \text{Old row} \\ \text{number} \end{matrix} - \left(\begin{matrix} \text{corresponding number} \\ \text{in old key row} \end{matrix} \times \begin{matrix} \text{corresponding} \\ \text{fixed ratio} \end{matrix} \right) = \text{new row number}$$

2,705	−	(445	×	1)	=	2,260
10.7	−	(0.7	×	1)	=	10
5	−	(1	×	1)	=	4
2	−	(2	×	1)	=	0
1	−	(0	×	1)	=	1
0	−	(0	×	1)	=	0
0	−	(1	×	1)	=	−1

Similarly, the new row S_2 is derived as follows (corresponding fixed ratio = 4/2 = 2):

$$\begin{matrix} \text{Old row} \\ \text{number} \end{matrix} - \left(\begin{matrix} \text{corresponding number} \\ \text{in old key row} \end{matrix} \times \begin{matrix} \text{corresponding} \\ \text{fixed ratio} \end{matrix} \right) = \text{new row number}$$

2,210	−	(445	×	2)	=	1,320
5.4	−	(0.7	×	2)	=	4
10	−	(1	×	2)	=	8
4	−	(2	×	2)	=	0
0	−	(0	×	2)	=	0
1	−	(0	×	2)	=	1
0	−	(1	×	2)	=	−2

These results are now entered in Tableau 6.II.

Tableau 6.II

Program	Profit per unit	Quantity	10 X	15 Y	20 Z	0 S_1	0 S_2	0 S_3	
S_1	0	2,260	10	4	0	1	0	−1	$\frac{2,260}{4} = 565$
S_2	0	1,320	4	8	0	0	1	−2	$\frac{1,320}{8} = 165 \checkmark$
Z	20	222.5	0.35	0.5	1	0	0	0.5	$\frac{222.5}{0.5} = 445$
Net-evaluation row:			3.00	5.00	0	0	0	−10	

Outgoing product   Incoming product

As indicated in Tableau 6.II, our second program calls for the production of $S_1 = 2,260$, $S_2 = 1,320$, and $Z = 222.5$ units. The variables S_3,

X, and Y are not in the solution (program) and, therefore, assume values of zero. The total profit contribution resulting from this program is $4,450 [2,260(0) + 1,320(0) + 222.5(20)].

6.7 DESIGNING ANOTHER PROGRAM (REVISION OF THE SECOND PROGRAM)

As the net-evaluation row* of Tableau 6.II shows, we still have two separate positive opportunity costs associated with *not* having 1 unit of each of variables (products) X and Y in the program. Hence, program 2, contained in Tableau 6.II, is not an optimum program. This calls for designing a new program and, therefore, deriving a new Tableau 6.III. The procedure for deriving Tableau 6.III is exactly the same as was followed in deriving Tableau 6.II. Before deriving Tableau 6.III, let us emphasize again that the numbers contained in the rows of Tableau 6.III will be the same as the coefficients of the different variables in Equations (21) to (23) in the systematic trial-and-error method.† Similarly, the numbers contained in the rows of Tableau 6.II are the coefficients of the different variables in Equations (17) to (19) of Chapter 3. This type of correspondence can be identified not only between the simplex method and the systematic trial-and-error method but essentially among all the methods of solving linear-programming problems discussed in this book. The advantage of the simplex method lies in the easy mechanical nature of its solution process, whereby a finite number of iterations takes us to the optimum solution. Important as the simplex method is for learning

* Calculated from

X	Y	Z	S_1	S_2	S_3
10	15	20	0	0	0
-10×0	-4×0	-0×0	-1×0	-0×0	$-(-1) \times 0$
-4×0	-8×0	-0×0	-0×0	-1×0	$-(-2) \times 0$
-0.35×20	-0.5×20	-1×20	-0×20	-0×20	-0.5×20
$+3.00$	$+5.00$	0	0	0	-10

† Rearrange the equations so that the constant terms are on the left-hand sides. See first footnote in Section 6.6.

purposes, we can really appreciate its value by noting that this type of procedure can easily be programmed into a computer. Thus, linear-programming problems of even very large dimensions can be solved in relatively short periods of time.

Derivation of Tableau 6.III

The two positive numbers in the net-evaluation row of Tableau 6.II, as noted earlier, indicate the existence of opportunity costs which must be eliminated. Since the introduction of each unit of Y into the program (at this stage) adds the highest profit contribution (or reduces the largest opportunity cost per unit), it is selected to be the "incoming" product. Thus, the column labeled Y is the key column. As before, we must now determine the limit on the quantity of Y which can be introduced into the program and thus identify the key row. The limiting quantities of variables S_1, S_2, and Z are shown on the right-hand side of Tableau 6.II; they indicate that row S_2 is the key row, and 8 is the key number.

By the rules of transformation presented earlier, the second row of Tableau 6.III (row Y) is derived from the second row of Tableau 6.II (row S_2) by simply dividing all the numbers in row S_2 by 8 (the key number). Row Y of Tableau 6.III becomes

165	0.5	1	0	0	0.125	-0.25

The nonkey rows of Tableau 6.II are transformed as follows:

Old row number

$$- \left(\begin{matrix} \text{corresponding number} \\ \text{in key row} \end{matrix} \times \begin{matrix} \text{corresponding} \\ \text{fixed ratio} \end{matrix} \right) = \text{new row number}$$

Calculations for row S_1 (fixed ratio = 4/8):

$$
\begin{aligned}
2{,}260 - (1{,}320 \times 0.5) &= 1{,}600 \\
10 - (4 \quad \times 0.5) &= 8 \\
4 - (8 \quad \times 0.5) &= 0 \\
0 - (0 \quad \times 0.5) &= 0 \\
1 - (0 \quad \times 0.5) &= 1 \\
0 - (1 \quad \times 0.5) &= -0.5 \\
-1 - (-2 \quad \times 0.5) &= 0
\end{aligned}
$$

Calculations for row Z (fixed ratio $= 0.5/8$):

$$222.5 \;-\; (1{,}320 \;\times\; 0.0625) \;=\; 140$$
$$0.35 \;-\; (4 \quad\times\; 0.0625) \;=\; 0.1$$
$$0.5 \;-\; (8 \quad\times\; 0.0625) \;=\; 0$$
$$1 \;-\; (0 \quad\times\; 0.0625) \;=\; 1$$
$$0 \;-\; (0 \quad\times\; 0.0625) \;=\; 0$$
$$0 \;-\; (1 \quad\times\; 0.0625) \;=\; -0.062$$
$$0.5 \;-\; (-2 \;\times\; 0.0625) \;=\; 0.625$$

These calculations are entered in Tableau 6.III. As indicated in Tableau 6.III, our third program calls for the production of $S_1 = 1{,}600$, $Y = 165$,

Tableau 6.III

	Profit per unit	Quantity	10 X	15 Y	20 Z	0 S_1	0 S_2	0 S_3	
S_1	0	1,600	8	0	0	1	−0.5	0	200✓
Y	15	165	0.5	1	0	0	0.125	−0.25	330
Z	20	140	0.1	0	1	0	−0.062	0.625	1,400
Net-evaluation row:			+0.5	0	0	0	−0.625	−8.75	

Outgoing product   Incoming product

and $Z = 140$. The variables S_2, S_3, and X are not in the program and, therefore, assume values of zero. The total profit contribution resulting from this program is \$5,275 $[1{,}600(0) + 165(15) + 140(20)]$.

6.8 DESIGNING ANOTHER PROGRAM (REVISION OF THE THIRD PROGRAM)

The net-evaluation row of Tableau 6.III still contains one positive number, indicating the existence of a positive opportunity cost. In order to eliminate this positive opportunity cost, variable (product) X should now be brought into the program. The column labeled X is therefore the key column.

The calculations for the limiting quantities of X that can be produced in view of the current program are shown at the extreme right-hand side of Tableau 6.III.   These calculations reveal that row S_1 is the key row. At the intersection of the key column and the key row is our key number, 8.   We are now in a position to determine the revised program.

As before, the derivation of the revised program (Tableau 6.IV) is accomplished by following the two transformation rules presented earlier. The key row of Tableau 6.III (row S_1) is transformed into row X of Tableau 6.IV, whereas the nonkey rows (rows Y and Z) are transformed into the corresponding rows of Tableau 6.IV.   The reader is encouraged to check these transformations, which are embodied in Tableau 6.IV. As indicated, our fourth program calls for the production of $X = 200$, $Y = 65$, and $Z = 120$ units.   The variables S_1, S_2, and S_3 are not in the

Tableau 6.IV

Program	Profit	Quantity per unit	10 X	15 Y	20 Z	0 S_1	0 S_2	0 S_3
X	10	200	1	0	0	0.125	-0.062	0
Y	15	65	0	1	0	-0.062	0.156	-0.25
Z	20	120	0	0	1	-0.012	-0.056	0.625
Net-evaluation row:			0	0	0	-0.062	-0.593	-8.75

solution (program) and, therefore, assume values of zero.* In other words, the fourth program will fully utilize the capacities of our cutting, folding, and packaging departments.   The total profit contribution resulting from this program is \$5,375 $[200(10) + 65(15) + 120(20)]$. The next question is: Is this our optimal program?

6.9   THE OPTIMAL PROGRAM

All the values in the net-evaluation row of Tableau 6.IV are either zero or negative.   This fact, as we established earlier, indicates that an optimal program has been obtained.

The economic interpretation of the entries in the net-evaluation row

* See the first column of Figure 6.1.

of the optimal program is interesting. Since the net evaluation of S_1 at this stage is -0.062, the introduction of 1 unit of S_1 (letting 1 unit of the cutting-department capacity stay idle) will decrease the objective function by $0.062. By the same reasoning, if we had one more unit of S_1, the objective function could be increased by $0.062. In other words, $0.062 gives us the *accounting* price or *shadow* price of 1 unit of cutting

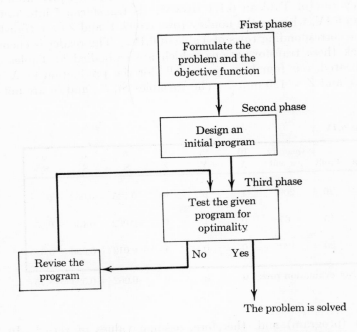

Figure 6.2 Schematic representation of the simplex algorithm.

capacity. It is similar to the economist's concept of the "worth" of a marginal unit of cutting-department capacity. Similarly, the shadow or accounting prices of S_2 and S_3 are, respectively, $0.593 and $8.75.* The value of all the available resources, using these shadow prices, can be calculated by multiplying the existing capacities of the different resources by their respective shadow prices and adding the products. In our case, this value is equal to $5,372 [2,705(0.062) + 2,210(0.593) + 445(8.75)]. Comparing this imputed value of the available resources with the value of

---

* The same information can be obtained by examining the *final* form of the objective function in the systematic trial-and-error method (see end of Section 3.3).

the objective function in the optimal program (Tableau 6.IV), we find that their magnitudes differ only by $3. Actually, the two values should have been exactly equal. The discrepancy has been caused by rounding errors. The fact that the value of the objective function in the optimal program equals the imputed value of the available resources has been called the *fundamental theorem of linear programming*. It is this theorem which is embodied in the concept of the *dual*. A discussion regarding the relationship between the so-called *primal* problem in linear programming and its dual will be presented in Chapter 8.

The solution of a linear-programming problem by the simplex method, as we have seen above, rests on a simple procedure consisting essentially of three phases. The first phase, of course, is to formulate the problem and the objective function. The second phase involves the design of an initial program which includes only the slack variables. The third phase consists in applying a test of optimality to determine whether a given program can be improved. The mechanics of the third phase, which repeats itself until an optimal solution (if it exists) is obtained, consists of two parts: (1) testing the optimality of the current program and (2) revising the current program, if necessary, according to definite rules of transformation for the key row and the nonkey rows of the simplex tableau containing the current program. A schematic diagram of this iterative procedure is given in Figure 6.2.

A step-by-step summary of the simplex procedure is given below.

6.10 SUMMARY PROCEDURE FOR THE SIMPLEX METHOD (MAXIMIZATION CASE)

Step 1 Formulate the Problem

a. Translate the technical specifications of the problem into inequalities, and make a precise statement of the objective function.

b. Convert the inequalities into equalities by the addition of nonnegative slack variables.* These equalities should be symmetric or balanced so that each slack variable appears in each equation with a proper coefficient.

c. Modify the objective function to include the slack variables.

---

* We assume here that all the inequalities are of the "less than or equal to" type. For a complete summary procedure for handling the three types of requirements ($\leq$, $\geq$, and $=$) see Section 7.8 of this book. Of course, in a maximization problem, it is convenient to convert all inequalities of the $\geq$ type into inequalities of the $\leq$ type by multiplying both sides of the inequalities by -1.

Step 2   Design an Initial Program (a Basic Feasible Solution)

Design the first program so that only the slack variables are included in the solution.   Place this program in a simplex tableau.   In the objective row, above each column variable, place the corresponding coefficient of that variable from step 1c.

Step 3   Test and Revise the Program

a. *Calculate the net-evaluation row.*   To get a number in the net-evaluation row under a column, multiply the entries in that column by the corresponding numbers in the objective column, and add the products. Then subtract this sum from the number listed in the objective row at the top of the column.   Enter the result in the net-evaluation row under the column.

b. *Test.*   Examine the entries in the net-evaluation row for the given simplex tableau.   If all the entries are zero or negative, the optimal solution has been obtained.   Otherwise, the presence of any positive entry in the net-evaluation row indicates that a better program can be obtained.

c. *Revise the program*
   1. Find the key column.   The column under which falls the largest positive net-evaluation-row entry is the key column.
   2. Find the key row and the key number.   Divide the entries in the "quantity" column by the corresponding *nonnegative* entries* of the key column to form *replacement ratios*, and compare these ratios. The row in which falls the *smallest* replacement ratio is the key row. The number which lies at the intersection of the key row and the key column is the key number.

---

* During any solution stage, the entries in the key column of a particular simplex tableau can present us with three situations.   First, one of the nonnegative entries may form the smallest replacement ratio, and definitely identify the key row.   In this case, the simplex algorithm (this section) is applied in a straightforward manner. Second, the nonnegative entries may be such that either the minimum replacement ratio is zero or there is a "tie" between two or more minimum replacement ratios. In such a case, the problem becomes *degenerate;* however, the simplex method may be continued according to rules discussed in Chapter 9.   Third, it may happen that all the entries in the key column are either zero or negative.   In this case, the incoming variable can be introduced into the program without any limit, and no current basis variable can be removed from the solution.   On the other hand, the values of the basis variables either remain the same or increase in magnitude without any limit (since all ratios of substitution are either zero or negative).   Thus, in the last case, we have a situation in which there are usually $m + 1$ variables in the solution, and the objective function usually has no finite maximum.

3. Transform the key row. Divide all the numbers in the key row (starting with and to the right of the "quantity" column) by the key number. The resulting numbers form the corresponding row of the next tableau.

4. Transform the nonkey rows. Subtract from the old row number of a given nonkey row (in each column) the product of the corresponding key-row number and the corresponding fixed ratio formed by dividing the old row number in the key column by the key number. The result will give the corresponding new row number. Make this transformation for all the nonkey rows.

5. Enter the results of (*3*) and (*4*) above in a tableau representing the revised program.

Step 4   Obtain the Optimal Program

Repeat steps 3 and 4 until an optimal program has been derived.

Linear-programming problems involving the minimization of an objective function usually contain structural constraints of the "greater than or equal to" type. They can also be solved by the simplex method. The simplex procedure for solving a linear-programming problem in which the objective is to minimize rather than maximize a given function, although basically the same as above, requires sufficient modifications to deserve the listing of a separate summary. This will be done after the simplex solution of a minimization problem is illustrated in the next chapter.

The Simplex Method. II

7.1 THE PROBLEM

The minimization case will be illustrated with a problem similar to the famous diet problem.* Let us formulate a hypothetical problem in which a person requires a certain amount of each of two vitamins per day.

Table 7.1

Vitamin	Food		Daily requirement
	F_1	F_2	
V	2	4	40
W	3	2	50
Cost per unit of food, cents.....	3	2.5	

The vitamins, V and W, are found in two different foods, F_1 and F_2. The amount of vitamin in each of the two foods, the respective prices per unit of each food, and the daily vitamin requirements are given in Table 7.1. The data indicate that 1 unit of F_1 (say 1 pound) contains 2 units of vitamin V and 3 units of vitamin W. Similarly, 1 unit of F_2 contains 4 units of vitamin V and 2 units of vitamin W. The daily requirement for vitamin V is *at least* 40 units, and for vitamin W *at least*

* Robert Dorfman, Paul A. Samuelson, and Robert M. Solow, "Linear Programming and Economic Analysis," p. 9, McGraw-Hill Book Company, New York, 1958.

130

50 units.* These, then, are the "technical" specifications of the problem. Our objective is to determine optimal quantities of foods F_1 and F_2 to be purchased so that the daily vitamin requirements are met and, simultaneously, the cost of purchasing the foods is minimized. Assuming that a represents the quantity of food F_1 and b the quantity of food F_2 to be purchased, we may state the problem in algebraic terms as follows:

Minimize (cost)

$$3a + 2.5b$$

subject to

$$2a + 4b \geq 40$$
$$3a + 2b \geq 50$$

and $a \geq 0, b \geq 0$.

7.2   SOLVING THE MINIMIZATION PROBLEM

As opposed to those in the problem discussed in the last chapter, the structural constraints here are of the "greater than or equal to" type. Hence, converting the inequalities into equalities requires the subtraction rather than the addition of "slack" variables. Let variables p and q represent, respectively, the quantity of vitamin V in excess of 40 units and the quantity of vitamin W in excess of 50 units. The introduction of these slack variables converts the above inequalities into the following equations:

$$2a + 4b - p = 40 \tag{1}$$
$$3a + 2b - q = 50 \tag{2}$$

A physical interpretation of these slack variables is somewhat parallel to the one given the slack variables in the maximization problem. Variables p and q can be thought of as "giveaway" or "free" products with the property that 1 unit of p contains 1 unit of vitamin V and 1 unit of q contains 1 unit of vitamin W. Obviously, we do not have to buy these giveaway products. If they exist in the program, they come as a "free

---

* Any intake of vitamins in excess of the daily requirements is assumed not to be harmful.

package"; p comes when we buy foods F_1 and F_2 with the purpose of satisfying the requirement for vitamin V, and q appears when we buy foods F_1 and F_2 to satisfy the requirement for vitamin W. Thus, if a specific program of purchasing foods F_1 and F_2 is such that the presence of p and q is required to satisfy Equations (1) and (2) above, then the *magnitudes of p and q will represent, respectively, the quantity of vitamin V in excess of* 40 *units and the quantity of vitamin W in excess of* 50 *units.*

With the above interpretation, it is easy to see that variables p and q are restricted to nonnegative values, and each has a cost coefficient of zero. A complete statement of the problem, therefore, is:

Minimize (cost)

$$3a + 2.5b + 0p + 0q$$

subject to

$$2a + 4b - p = 40 \tag{3}$$
$$3a + 2b - q = 50 \tag{4}$$

and $a \geq 0$, $b \geq 0$, $p \geq 0$, $q \geq 0$.

Artificial Slack Variables

Let us examine Equation (3). If we let structural variables a and b equal zero as we did in designing the first program while solving the maximization problem by the simplex method, we obtain a value of -40 for the slack variable p. Any such negative value is unacceptable, since it violates the nonnegativity constraint for p and does not make any sense in terms of the physical interpretation of the slack variables. We face a similar difficulty in connection with Equation (4). Yet we know that the trivial initial solution in the simplex method, as we have seen in the last chapter, is always obtained by letting all the structural variables equal zero. We propose, therefore, at this stage, to modify the statement of our problem in such a way that we can make a and b equal to zero in the above equations and still have positive-valued slack variables satisfying these equations. This is accomplished by introducing into the original inequalities, in addition to the regular slack variables, the so-called "artificial" slack variables. The artificial slack variables will be represented by the capital letter A with proper subscript. Thus, we can modify Equations (3) and (4) with the addition of artificial slack variables

A_1 and A_2, respectively:*

$$2a + 4b - p + A_1 = 40$$
$$3a + 2b - q + A_2 = 50$$

In this problem, the artificial slack variables A_1 and A_2 can be thought of as "imaginary foods," each unit containing 1 unit of the pertinent vitamin. For example, we can assume here that 1 unit of A_1 contains 1 unit of vitamin V, whereas one unit of A_2 contains 1 unit of vitamin W. In this sense, A_1 is similar to p, and A_2 is similar to q. Also, both A_1 and A_2 are restricted to nonnegative values for obvious reasons. However, the correspondence between the slack and artificial slack variables does not hold in the matter of cost coefficients. Whereas slack variables have zeros as their cost coefficients, each artificial slack variable is assigned an infinitely large cost coefficient (usually denoted by M).

Thus, the addition of the slack and artificial slack variables converts the original problem to the following:

Minimize (cost)

$$3a + 2.5b + 0p + 0q + MA_1 + MA_2$$

subject to

$$2a + 4b - p + A_1 = 40 \tag{5}$$
$$3a + 2b - q + A_2 = 50 \tag{6}$$

and a, b, p, q, A_1, and $A_2 \geq 0$.

If in Equations (5) and (6) we let variables a, b, p, and q assume values of zero, the artificial slack variables A_1 and A_2 will have positive values. We can see, therefore, that the inclusion of artificial slack variables will permit us to design an initial program in which no units of foods F_1 and F_2 are purchased, and yet the nonnegativity constraints are not violated.†

In other words, these variables enable us to make a convenient and correct start in obtaining an optimal solution by the simplex method. Further, having attached to each artificial slack variable an extremely large cost coefficient M, we can be certain that these variables can never

---

* For each given constraint which has to be entered into the initial simplex tableau with the addition of an artificial slack variable, we need only one such artificial slack variable.

† The reader will recall that such an initial program, when applied to Equations (3) and (4), violated the nonnegativity constraints for variables p and q.

enter into the optimal solution.*   The inclusion of even 1 unit of an artificial slack variable in any program would result in a prohibitive cost.

The solution to our modified problem (modified by the inclusion of A_1, A_2, etc.) will therefore give us the solution to the original problem.

Balancing the Equations for the Initial Simplex Tableau

It will be recalled from the simplex solution of our maximization problem that the initial program was that which involved only the imaginary products S_1, S_2, etc.   For a two-dimensional problem, this corresponds to starting the solution at the origin in the graphical method and to expressing the requirement vector P_0 in terms of basis vectors $\begin{bmatrix} 1 \\ 0 \end{bmatrix}$ and $\begin{bmatrix} 0 \\ 1 \end{bmatrix}$ in the vector method.   In so far as our present minimization problem is two-dimensional, we know that its requirement vector $P_0 = \begin{bmatrix} 40 \\ 50 \end{bmatrix}$ can be expressed in terms of the basis vectors $\begin{bmatrix} 1 \\ 0 \end{bmatrix}$ and $\begin{bmatrix} 0 \\ 1 \end{bmatrix}$.   If we write Equations (5) and (6) in a balanced form, we immediately observe that the introduction of A_1 and A_2 has brought us to the point where basis vectors $\begin{bmatrix} 1 \\ 0 \end{bmatrix}$ and $\begin{bmatrix} 0 \\ 1 \end{bmatrix}$ become available, and the equations can be entered directly into the first simplex tableau [see Equations (7) and (8)].   In the balanced form, our problem can be stated as follows:

Minimize (cost)

$$3a + 2b + 0p + 0q + MA_1 + MA_2$$

subject to

$$2a + 4b - 1p + 0q + 1A_1 + 0A_2 = 40 \tag{7}$$

$$3a + 2b + 0p - 1q + 0A_1 + 1A_2 = 50 \tag{8}$$

and a, b, p, q, A_1, and $A_2 \geq 0$.

We recognize the above as a typical linear-programming problem, which will now be solved by the simplex method.

* If, in any linear-programming problem involving artificial slack variables, the application of the simplex method fails to remove all artificial slack variables from the solution basis, the original problem has no solution.

7.3 DESIGNING THE INITIAL PROGRAM

As discussed above, our first program is obtained by letting each of the variables a, b, p, and q assume a value of zero. This means [from Equations (7) and (8)] that the initial program calls for purchasing 40 units of A_1 and 50 units of A_2. This program is given in Tableau 7.I. This

Tableau 7.I

Program	Cost per unit	Quantity	3 a	2.5 b	0 p	0 q	M A_1	M A_2
A_1	M	40	2	4	-1	0	1	0
A_2	M	50	3	2	0	-1	0	1
Net-evaluation row:			$3-5M$	$\frac{5}{2}-6M$	M	M	0	0

Outgoing variable

Incoming variable

program, as the contents of Tableau 7.I show, involves a cost of $90M$ dollars—a prohibitive figure. Obviously, we can design a better program.

7.4 REVISION OF THE INITIAL OR CURRENT PROGRAM

As in the simplex method for solving a maximization problem, revision of the current program in the case of a minimization problem requires (1) the calculation of the net-evaluation row, (2) identification of the key column, (3) identification of the key row and the key number, and (4) transformation of the key row and the nonkey rows into the new tableau containing the revised program. The mechanics of these steps is exactly the same as in the maximization problem. However, in the minimization case it is the largest negative entry in the net-evaluation row (as opposed to the largest positive entry in the maximization case) which identifies the key column. The reason for this is obvious. In the minimization case, if the net-evaluation entry under a particular column variable is negative, it is indicative of the fact that the inclusion of this variable in the new basis (by replacement of one of the current basis variables) will decrease the value of the objective function.

To calculate the net-evaluation row, we determine the net effect on cost of including (purchasing), at this stage of the program, 1 unit of each of a, b, p, q, A_1, and A_2. For example, as the ratios of substitution of Tableau 7.I show, purchasing 1 unit of b (or producing 1 unit of b) means that the purchase of A_1 will have to be reduced by 4 units and, simultaneously, the purchase of A_2 will have to be reduced by 2 units.* The net effect of introducing 1 unit of b is to reduce the total cost by $2.5 - 6M$ cents $[+1(2.5) - 4M - 2M]$. In other words, replacing A_1 or A_2 with b, at this stage, reduces the total cost. Similar information can be obtained for other variables not included in the program at this stage. The net-evaluation row of Tableau 7.I contains this information.

The negatives of the entries in the net-evaluation row represent the *opportunity costs* of *not* having 1 unit of each of the variables in the program. For example, the opportunity cost of not having 1 unit of b in the solution is $-(2.5 - 6M) = 6M - 2.5$ cents, a positive opportunity cost which must be eliminated. Since a positive opportunity cost corresponds to a negative entry in the net-evaluation row in the case of a minimization problem, we can state the following decision rule for testing the optimality of a given program in a minimization problem: *So long as there exists even a single negative number in the net-evaluation row of a minimization problem, the optimal solution has not been obtained.*†

As the entries in the net-evaluation row of Tableau 7.I show, variables a and b are the only variables whose introduction to the program will reduce the total cost. We observe that the introduction of 1 unit of a and 1 unit of b into the program, at this stage, will reduce the objective function by $3 - 5M$ and $2.5 - 6M$ cents, respectively. Since the introduction of 1 unit of b reduces the cost more than does 1 unit of a, we should first bring b into the program.‡ The column labeled b is therefore the key column.

Next we must determine how many units of b can be purchased without making either A_1 or A_2 negative.

From row A_1, the maximum amount of b that can be brought into the solution is $40/4 = 10$ units.

---

* The reader may observe here the symmetry of interpretation between the simplex tableaux of maximization and minimization problems in linear programming.

† We assume that the net-evaluation row is calculated according to the procedure in Section 6.10.

‡ In so far as M represents a very large cost, the comparative magnitudes of the net-evaluation-row entries can be ascertained simply by comparing the number of M's. Thus, in Tableau 7.I, the net-evaluation entry $\frac{5}{2} - 6M$ is, in absolute terms, larger than $3 - 5M$. This means that, for reducing the cost function, 1 unit of b is preferable, at this stage, to 1 unit of a.

From row A_2, the maximum amount of b that can be brought into the solution is $50/2 = 25$ units.

We see, therefore, that row A_1 provides the limiting case. This is, in other words, our key row, and the key number is 4.

The rest of the procedure for revising the current program is exactly the same as that followed in the simplex method for a maximization problem. The key row (row A_1 in this case) is transformed by dividing all its entries by the key number; the nonkey row (row A_2 in this case) is transformed according to the transformation rule used in the last chapter.* Accordingly, a new program is designed. Tableau 7.II lists this program along

Tableau 7.II

Program	Cost per unit	Quantity	3 a	2.5 b	0 p	0 q	M A_1	M A_2
b	2.5	10	$\frac{1}{2}$	1	$-\frac{1}{4}$	0	$\frac{1}{4}$	0
A_2	M	30	2	0	$\frac{1}{2}$	-1	$-\frac{1}{2}$	1

Net-evaluation row: $\quad \frac{7}{4}-2M \quad$ 0 $\quad \frac{5}{8}-\frac{1}{2}M \quad M \quad -\frac{5}{8}+\frac{3}{2}M \quad$ 0

Outgoing variable   Incoming variable

with other pertinent information. The reader should verify the contents of Tableau 7.II and the subsequent programs in this chapter.

7.5 PROGRAM 3 (REVISION OF THE SECOND PROGRAM)

The net-evaluation row of Tableau 7.II has two negative entries. The presence of these negative entries means that the optimal program has not been obtained as yet, and hence our second program can be improved.

Since the net-evaluation entry under column a has a larger negative

---

* As the reader will recall, the transformation rule for the nonkey rows is

Old row number $- \left(\begin{matrix} \text{corresponding number} \\ \text{in key row} \end{matrix} \times \begin{matrix} \text{corresponding} \\ \text{fixed ratio} \end{matrix} \right) = $ new row number

where

Fixed ratio $= \dfrac{\text{old row number in key column}}{\text{key number}}$

value than that under column p, the variable a should be brought into the solution next. This is accomplished in precisely the same manner as before. Our revised program is given in Tableau 7.III.

Tableau 7.III

Program	Cost per unit	Quantity	3 a	2.5 b	0 p	0 q	M A_1	M A_2
b	2.5	$\frac{5}{2}$	0	1	$-\frac{3}{8}$	$\frac{1}{4}$	$\frac{3}{8}$	$-\frac{1}{4}$
a	3	15	1	0	$\frac{1}{4}$	$-\frac{1}{2}$	$-\frac{1}{4}$	$\frac{1}{2}$

	Net-evaluation row:	0	0	$\frac{3}{16}$	$\frac{7}{8}$	$M - \frac{3}{16}$	$M - \frac{7}{8}$

7.6 THE OPTIMAL PROGRAM

Since all the entries in the net-evaluation row of Tableau 7.III are positive, the optimal solution to our problem has been obtained. This optimal program assigns a value of 15 to variable a and $\frac{5}{2}$ to variable b. In other words, this optimal program calls for purchasing 15 units of food F_1 and $\frac{5}{2}$ units of food F_2 daily, with an attendant cost of 51.25 cents. As a quick check will show, this program meets the daily requirements of vitamins V and W.

7.7 SUMMARY PROCEDURE FOR THE SIMPLEX METHOD (MINIMIZATION CASE)

Step 1 *Formulate the Problem*

a. Translate the technical specification of the problem into inequalities, and make a precise statement of the objective function.
b. Convert the inequalities into equalities by the subtraction of nonnegative slack variables.* Then modify these equations by the addition of nonnegative artificial slack variables. These equations should

---

* We assume that all the inequalities are of the "greater than or equal to" type. For a complete summary procedure for handling the three types of requirements ($\leq$, $\geq$, $=$), see Section 7.8. Of course, in a minimization problem, it is convenient to convert all inequalities of the $\leq$ type into inequalities of the $\geq$ type by multiplying both sides of each inequality by -1.

be symmetric or balanced so that each slack and artificial slack variable appears in each equation with a proper coefficient.

c. Modify the objective function to include all the slack and artificial slack variables.

Step 2   *Design an Initial Program (a Basic Feasible Solution)*

Design the first program so that only the artificial slack variables are included in the solution. Place this program in a simplex tableau. In the objective row, above each column variable, place the corresponding coefficient of that variable from step 1c. In particular, place a zero above each column containing a slack variable, and an infinitely large number M above each column containing an artificial slack variable.

Step 3   *Test and Revise the Program*

a. Calculate the net-evaluation row. To get a number in the net-evaluation row under a column, multiply the entries in that column by the corresponding number in the objective column, and add the products. Then subtract this sum from the number listed in the objective row above the column. Enter the result in the net-evaluation row under the column.

b. Test. Examine the entries in the net-evaluation row for the given simplex tableau. If all the entries are zero or positive, the optimum solution has been obtained. Otherwise, the presence of any negative entry in the net-evaluation row indicates that a better program can be obtained.

c. Revise the program

   1. Find the key column. The column under which falls the largest negative net-evaluation-row entry is the key column.

   2. Find the key row and the key number. Divide the entries in the "Quantity" column by the corresponding nonnegative entries* in the key column to form replacement ratios, and compare these ratios. The row in which the *smallest* replacement ratio falls is the key row. The number which lies at the intersection of the key row and the key column is the key number.

   3. Transform the key row. Divide all the numbers in the key row (starting with and to the right of the "Quantity" column) by the key number. The resulting numbers form the corresponding row of the next tableau.

---

* See last footnote in Section 6.10.

4. Transform the nonkey rows. Subtract from the old row number of a given nonkey row (in each column) the product of the corresponding key row number and the corresponding fixed ratio formed by dividing the old row number in the key column by the key number. The result will give the corresponding new row number. Make this transformation for all the nonkey rows.

5. Enter the results of (*3*) and (*4*) above in a tableau representing the revised program.

Step 4   Obtain the Optimal Program

Repeat steps 3 and 4 until an optimal program has been derived.

We repeat the following comments comparing the maximization and minimization problems as solved by the simplex method.

The procedure for calculating the net-evaluation row is the same in both cases. However, whereas the largest positive value is chosen to identify the incoming product in a maximization problem, the most negative value is chosen in a minimization problem. The rest of the mechanics, namely, the transformation of the key and nonkey rows, is exactly the same. The decision rule identifying the optimal solution is the absence of any positive value in the net-evaluation row in the maximization problem, and the absence of any negative value in the minimization problem.

7.8   PROCEDURE FOR MODIFYING GIVEN STRUCTURAL CONSTRAINTS

The original technical specifications of a given linear-programming problem can be expressed by three different types of structural constraints. First, there are those constraints which, in their original form, can be represented by inequalities of the "less than or equal to" type. Second, there are constraints which, in their original form, can be represented by inequalities of the "greater than or equal to" type. These two types of constraints, consisting of inequalities in their original form, are modified by converting the inequalities to equalities. This is done, as the reader will recall, to arrange the data properly for the construction of the first simplex tableau. We have, however, not discussed so far the case in which the constraints, even in their original form, must be expressed by exact equations. These constraints, which form our third category, must also be modified for purposes of building the first simplex tableau. How do we accomplish this? Case III of the following summary will provide the answer. Illustrated in this summary are the mechanics for converting the various inequalities and equalities (expressing the constraints in their original form) into balanced equations for purposes of constructing the first simplex tableau.

Case I   Inequalities of the "Less than or Equal to" Type ($\leq$)

Each inequality of the "less than or equal to" type is converted to an equation by the addition of a nonnegative slack variable having a profit (objective) coefficient of zero.

Example

Maximize $10X + 15Y$ subject to

$$4X + 6Y \leq 60$$
$$3X + 4Y \leq 80$$

and $X \geq 0$, $Y \geq 0$.

The corresponding equations for the first simplex tableau are

$$4X + 6Y + 1S_1 + 0S_2 = 60$$
$$3X + 4Y + 0S_1 + 1S_2 = 80$$

and the modified objective function is:
    Maximize

$$10X + 15Y + 0S_1 + 0S_2$$

Case II   Inequalities of the "Greater than or Equal to" Type ($\geq$)

Each inequality of the "greater than or equal to" type is converted to an equation by first subtracting a nonnegative slack variable having a cost (objective) coefficient of zero and then adding a nonnegative artificial slack variable having a cost (objective) coefficient of M (an infinitely large number).

Example

Minimize $300a + 180b$ subject to

$8a + 5b \geq 80$

$4a + 2b \geq 70$

and $a \geq 0$, $b \geq 0$.

The corresponding equations for the first simplex tableau are

$8a + 5b - 1p + 0q + 1A_1 + 0A_2 = 80$

$4a + 2b + 0p - 1q + 0A_1 + 1A_2 = 70$

and the modified objective function is:
   Minimize

$300a + 180b + 0p + 0q + MA_1 + MA_2$

Case III   The Mixed Case

(In addition to inequalities, the given problem has equalities in the initial statement of the problem.)

In this case inequalities are handled as detailed above.   Each equality is modified by the addition of a nonnegative artificial slack variable.

Example

Minimize $7a + 15b$ subject to

$2a + 4b \geq 20$

$5a + 8b = 30$

The corresponding equations for the first simplex tableau are

$2a + 4b - 1p + 1A_1 + 0A_2 = 20$

$5a + 8b + 0p + 0A_1 + 1A_2 = 30$

and the modified objective function is:
   Minimize

$7a + 15b + 0p + MA_1 + MA_2$

8.1 THE DUAL PROBLEM AND ITS SOLUTION

Associated with every linear-programming problem is its dual. The meaning of the dual will become clear in the following discussion dealing with the vitamin problem solved earlier (Table 7.1). The data of the problem are reproduced in Table 8.1.

Table 8.1

Vitamin	Food		Daily requirement
	F_1	F_2	
V	2	4	40
W	3	2	50
Cost per unit of food, cents.....	3	2.5	

Let us consider that the foods F_1 and F_2, among other things, are being sold by a corner grocery store. The owner of the store realizes that foods F_1 and F_2 have a market value because of their vitamin contents V and W. His problem is to determine selling prices, say X cents per unit of vitamin V, and Y cents per unit of vitamin W. He realizes that his unit prices for the vitamins should be assigned in such a manner that the resultant selling prices of the two foods are less than or equal to the market prices. In other words, X and Y are to be assigned values in such a manner that the computed costs for foods F_1 and F_2 are less than or

143

equal to 3 and 2.5 cents per unit, respectively. Otherwise, since the market is assumed to be competitive, he will have to price the foods F_1 and F_2 at more than 3 and 2.5 cents per unit and hence will lose his customers. At the same time, he would like to maximize his return, which is given by $40X + 50Y$ (since the daily vitamin requirement is 40 units of vitamin V and 50 units of vitamin W). Let us state the grocer's problem in terms of inequalities and an objective function:

Maximize the objective function $40X + 50Y$ subject to

$$2X + 3Y \leq 3$$
$$4X + 2Y \leq 2.5 \tag{1}$$

and $X \geq 0$, $Y \geq 0$.

The above statement of the grocer's problem is the *dual* to the original problem stated in Table 7.1. For purposes of identification, the original

Tableau 8.I

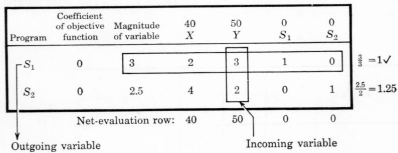

Program	Coefficient of objective function	Magnitude of variable	40 X	50 Y	0 S_1	0 S_2	
S_1	0	3	2	3	1	0	$\frac{3}{3} = 1\checkmark$
S_2	0	2.5	4	2	0	1	$\frac{2.5}{2} = 1.25$
	Net-evaluation row:		40	50	0	0	

Outgoing variable   Incoming variable

problem is called the *primal* problem. Of course, if we first state the problem as (1) above, then (1) is the primal problem, and the problem in Table 7.1 is its dual. The point to emphasize here is that for every linear-programming problem there is a unique dual problem.

The reader will realize that the grocer's problem is a typical linear-programming problem in which the objective function is to be maximized. This problem can easily be solved by the simplex method. The various stages of the simplex solution are summarized in Tableaux 8.I to 8.III.

Since all the entries in the net-evaluation row in Tableau 8.III are either zero or negative, the optimal solution to the maximization problem has been obtained. The grocer has, in other words, placed a "worth" of $\frac{3}{16}$ cent per unit on vitamin V, and $\frac{7}{8}$ cent per unit on vitamin W. The value of the objective function is $50 \times \frac{7}{8} + 40 \times \frac{3}{16} = 51.25$ cents, which is exactly the same as was obtained in the minimization problem concerned with purchasing the foods F_1 and F_2 (see Section 7.6).

Tableau 8.II

Program	Coefficient of objective function	Magnitude of variable	40 X	50 Y	0 S_1	0 S_2	
Y	50	1	$\frac{2}{3}$	1	$\frac{1}{3}$	0	$\frac{1}{\frac{2}{3}} = \frac{3}{2}$
S_2	0	$\frac{1}{2}$	$\frac{8}{3}$	0	$-\frac{2}{3}$	1	$\frac{\frac{1}{2}}{\frac{8}{3}} = \frac{3}{16}$ ✓
	Net-evaluation row:		$6\frac{2}{3}$	0	$-16\frac{2}{3}$	0	

Outgoing variable   Incoming variable

Tableau 8.III

Program	Coefficient of objective function	Magnitude of variable	40 X	50 Y	0 S_1	0 S_2
Y	50	$\frac{7}{8}$	0	1	$\frac{1}{2}$	$-\frac{1}{4}$
X	40	$\frac{3}{16}$	1	0	$-\frac{1}{4}$	$\frac{3}{8}$
	Net-evaluation row:	0	0	-15	$-\frac{5}{2}$	

8.2 A COMPARISON OF THE OPTIMAL TABLEAUX OF THE PRIMAL PROBLEM AND ITS DUAL

Let us now place the optimal tableau of the primal problem concerned with purchasing foods F_1 and F_2 (Tableau 7.III) next to the optimal tableau of its dual (Tableau 8.III) and make some important observations (see Figure 8.1).

The objective functions of the two optimal tableaux assume identical values. We note further that entries (with signs changed) in the net-evaluation row under columns S_1 and S_2 of the optimal tableau of the dual are the same as entries under the "Quantity" column in the optimal tableau of the primal problem. Also, the magnitudes of the variables X and Y are exactly the same as the entries in the net-evaluation row under

146

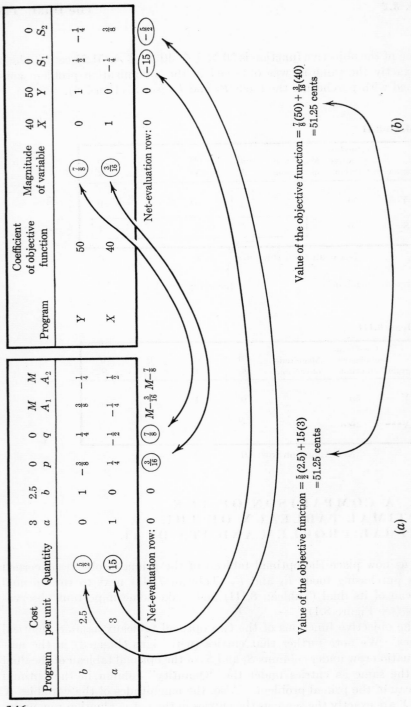

Figure 8.1 (a) Optimal tableau of the primal problem; (b) optimal tableau of the dual.

(a)

Program	Cost per unit	Quantity	3 a	2.5 b	0 p	0 q	M A₁	M A₂
b	2.5	$\frac{5}{2}$	0	1	$-\frac{3}{8}$	$\frac{1}{4}$	$\frac{3}{8}$	$-\frac{1}{4}$
a	3	15	1	0	$\frac{1}{4}$	$-\frac{1}{2}$	$-\frac{1}{4}$	$\frac{1}{2}$
	Net-evaluation row:	0	0	0	$\frac{3}{16}$	$\frac{7}{8}$	$M-\frac{3}{16}$	$M-\frac{7}{8}$

Value of the objective function $= \frac{5}{2}(2.5) + 15(3)$
$= 51.25$ cents

(b)

Program	Coefficient of objective function	Magnitude of variable	40 X	50 Y	0 S₁	0 S₂
Y	50	$\frac{7}{8}$	0	1	$\frac{1}{8}$	$-\frac{1}{4}$
X	40	$\frac{3}{16}$	1	0	$-\frac{1}{4}$	$\frac{3}{8}$
	Net-evaluation row:	0	0	0	−15	$-\frac{5}{2}$

Value of the objective function $= \frac{7}{8}(50) + \frac{3}{16}(40)$
$= 51.25$ cents

columns p and q of the optimal tableau of the primal problem.   This type of correspondence between the optimal tableaux of the primal and its dual always exists.   Thus, the solution to a primal problem in linear programming can always provide a solution to its dual.   The relationships just described should be carefully checked and grasped by the reader.

8.3 SYMMETRY BETWEEN THE PRIMAL AND ITS DUAL

The symmetry between our primal problem and its dual is summarized below:

$$\begin{array}{c}
\text{Maximize:} \\
\begin{array}{cccc}
2a & + & 4b & \geq & 40 \\
X & & X & & X \\
+ & & + & & + \\
3a & + & 2b & \geq & 50 \\
Y & & Y & & Y \\
\wedge\!| & & \wedge\!| & & \\
\end{array}
\end{array}$$

Minimize: $3a + 2.5b$

Reading horizontally, we have the minimization problem as stated in Chapter 7.   Reading vertically, we have the dual (in this case a maximization problem) as stated in Section 8.1.

Let us now consider the maximization problem (Table 6.1) which we solved by the simplex method in Chapter 6.   This problem also has its dual.   We shall first form the dual and then explain its economic significance.

$$\begin{array}{c}
\text{Minimize:} \\
\begin{array}{cccccc}
10.7X & + & 5Y & + & 2Z & \leq & 2{,}705 \\
a & & a & & a & & a \\
+ & & + & & + & & + \\
5.4X & + & 10Y & + & 4Z & \leq & 2{,}210 \\
b & & b & & b & & b \\
+ & & + & & + & & + \\
0.7X & + & 1Y & + & 2Z & \leq & 445 \\
c & & c & & c & & c \\
\vee\!| & & \vee\!| & & \vee\!| & & \\
\end{array}
\end{array}$$

Maximize: $10X + 15Y + 20Z$

Reading horizontally, we have listed the maximization problem of Table 6.1. Reading vertically, its dual can be stated as follows:

$$10.7a + 5.4b + 0.7c \geq 10$$

$$5a + 10b + 1c \geq 15$$

$$2a + 4b + 2c \geq 20$$

The objective function to be minimized is

$$2,705a + 2,210b + 445c$$

In this case, the concept of the dual can be explained in the following terms. The management of the company producing products X, Y, and Z realizes that the per-unit profit contributions of these products are $10, $15, and $20, respectively. Further, they know that technical specifications, as given in Table 6.1, have fixed the relative proportions of the "resources" required to produce these products. For example, to produce 1 unit of product X, we need 10.7 units of cutting-department capacity, 5.4 units of folding-department capacity, and 0.7 unit of packaging-department capacity. Thus, for each product, a relative expenditure of these resources brings the company certain contributions to profit. Since a profit-seeking organization is always searching for alternative uses for its resources, the management wishes to get some idea of the worth of these resources. Let us assume that the worths in alternative uses per unit of cutting, folding, and packaging capacity are a, b, and c dollars, respectively. Whatever values are assigned to variables a, b, and c, the worth of 1 unit of product X is $10.7a + 5.4b + 0.7c$, for 10.7, 5.4, and 0.7 are the relative numbers of units of the three resources required to produce 1 unit of product X. Then, as a first formulation, in the case of product X, the management will want to know whether $10.7a + 5.4b + 0.7c$ is greater than or equal to the $10 per unit profit contribution of X. If it is greater than $10, the production of product X must be stopped and resources directed to alternative uses. If it is equal to $10, then the production of X can be continued, since alternative uses will not bring any greater contribution to profit. Furthermore, since we are dealing with interdependent activities, the same type of relationships must simultaneously hold for the other products (Y and Z) considered in this problem. This, then, is the reason for writing the three inequalities of the dual problem.

The worth of alternative uses is sometimes called the *shadow* or *implicit* or *accounting* price of these resources. The specific values of these shadow prices are determined along with a minimization objective, which was stated in the dual problem above. Intuitively, one can think of this as minimizing the expenditure associated with the existing capacities of the various resources. Thus, in this problem, the objective is to minimize $2,705a + 2,210b + 445c$.

The specific solution of this dual problem can be gained, as explained earlier, by inspecting the optimal simplex tableau of the primal problem. Thus, by reading the entries (with signs changed) in the net-evaluation row under columns S_1, S_2, and S_3 of the optimal Tableau 6.IV, we determine immediately that $a = 0.062$, $b = 0.593$, and $c = 8.75$. In other words, one unit of cutting-department capacity, in the present situation, is worth \$0.062 to the company, and the shadow prices per unit of folding and packaging capacities are \$0.593 and \$8.75, respectively.

Substitution of these values for a, b, and c in the inequalities of the dual problem shows that the constraints are not violated. Further, the value of the objective function of the dual is \$5,372, which differs from the value of the objective function in the primal problem by only \$3. The discrepancy has been caused by rounding errors. Otherwise, as we know, the magnitudes of the objective functions of the primal problem and its dual are exactly the same.

8.4   SOME COMMENTS ON THE DUAL

It should be noted that if there are n structural variables and m slack variables in the primal problem, there will be m structural variables and n slack variables in its dual. That is, the transpose of the matrix (of the structural coefficients) of the primal problem gives the matrix (of the structural coefficients) of the dual. This point is perhaps not clear, since in both examples used to illustrate the dual in this chapter the number of structural variables was the same as the number of slack variables. However, the reader can easily verify this relationship between the numbers of structural and slack variables of the primal and its dual by formulating a primal problem in which the number of structural variables is more than or less than the number of slack variables.

The reader may very well wonder about the intrinsic value of the dual in view of our statement that the optimal solution of the primal contains, in itself, the solution of the dual. Knowledge of the dual is important for two main reasons. First, the economic interpretation of the dual is useful in making future decisions in the activities being programmed. Second, the solution of a linear-programming problem may be easier to obtain through the dual than through the primal problem. This is true of cases in which, in the primal problem, the number of structural variables is considerably less than the number of slack variables. For example, consider a primal problem involving two structural variables and seven slack variables. The initial simplex tableau for this problem will have seven rows. On the other hand, the same problem can be solved via its dual, for which the initial simplex tableau will have only two rows.

Degeneracy

9.1 INTRODUCTION

It will be recalled that the simplex method is based on a set of rules whereby we proceed from one basic feasible solution to the next until an optimal solution, if it exists, is obtained. Each new simplex program, if we think in terms of vectors, is obtained by choosing a new set of basis vectors. The iterative process, therefore, consists in going from the old basis to a new basis. The new basis is chosen by replacing at least one of the vectors currently in the basis with only *one* of the nonbasis vectors. The vector to be introduced and the vector to be replaced correspond, respectively, to the key column and the key row of the simplex method. To proceed from one solution to the next by the simplex method, as the reader will recall from Chapter 6, requires the identification of the key column and the key row.

Selection of the key column is a simple task, for it simply involves the identification of the column containing the largest positive entry (maximization case) or the largest negative entry (minimization case) in the net-evaluation row of a simplex tableau. However, in selecting the key row for purposes of replacing one of the basis vectors, we can face two difficulties.

1. The *initial* simplex tableau may be such that one or more of the variables currently in the basis has a value of zero (one or more entries in the "Quantity" column is zero). If this happens, the minimum replacement ratio will be zero. It will, then, appear that the replacement process cannot be continued, for the variable to be replaced is *already* zero.

2. The minimum nonnegative replacement ratios* for two or more

* See Section 6.5.

variables currently in the basis may be the same.   If this happens, there is a "tie" in terms of selection of the key row.   In this case, removal of one of the tied variables will also reduce the other tied variable(s) to zero. In the next tableau of this case, therefore, one or more of the basis vectors will have a value of zero.

Both the above conditions give rise to a phenomenon known as *degeneracy*.   Attempts to solve a degenerate linear-programming problem will show that either (1) after a finite number of iterations the optimum solution can be obtained or (2) the problem begins to cycle,* thereby preventing the attainment of the optimum solution.   We shall comment on these two aspects of degeneracy after solving the following linear-programming problem.

9.2   THE PROBLEM

Maximize $22X + 30Y + 25Z$ subject to

$$2X + 2Y \quad\quad \leq 100$$

$$2X + Y + Z \leq 100$$

$$X + 2Y + 2Z \leq 100$$

and $X \geq 0,\ Y \geq 0,\ Z \geq 0$.

Tableau 9.1a

Program	Profit per unit	Quantity	22 X	30 Y	25 Z	0 S_1	0 S_2	0 S_3	
S_1	0	100	2	2	0	1	0	0	$\frac{100}{2} = 50$
S_2	0	100	2	1	1	0	1	0	$\frac{100}{1} = 100$
S_3	0	100	1	2	2	0	0	1	$\frac{100}{2} = 50$
Net-evaluation row:			22	30	25	0	0	0	

Putting the above problem in the initial simplex tableau, we get Tableau 9.1a.   Calculation of the net-evaluation row in Tableau 9.1a

---

* That is, during the solution stages, we keep returning to the same basis.

shows that column Y is our key column. As previously, our next task is to choose a key row by identifying that variable (product) which is to be replaced by the incoming variable or product (in this case the incoming variable is Y). But there is no unique key row in Tableau 9.Ia, since both row S_1 and row S_3 provide the limiting case. In other words, we have a tie between row S_1 and row S_3.

The introduction of 50 units of Y at this stage will require removing all units of S_1 and S_3 from the solution. This means that our next program will consist of 50 units of Y and 50 units of S_2. It appears, therefore, that our next tableau would have only two rows instead of the three rows contained in Tableau 9.Ia. This situation, as the reader will quickly realize, is different from anything that we have encountered so far. We have never faced a situation in which more than one variable at a time had to be replaced from a given basis. In every linear-programming problem discussed earlier, all simplex tableaux, during all the solution stages, had the same number of rows. How, then, do we proceed in the case in which a tie appears? The answer is that, in so far as the simplex method requires the replacement of only one basis variable at a time, we should somehow break the tie between row S_1 and row S_3 by designating one of them as the key row. Then, only that variable which falls in the key row should be removed from the basis. The mechanics for accomplishing this is discussed below.

9.3 SOLVING A DEGENERATE PROBLEM

In Tableau 9.Ia we have encountered a degenerate situation. How do we resolve the degeneracy? Some rule is obviously needed to break the tie between the two variables S_1 and S_3. Several arbitrary rules have been suggested for making this decision. One such rule is that the variable whose subscript is smallest* should be removed first. Another rule calls for removing that variable whose subscript is found first in a table of random numbers. Another alternative, of course, is to remove one of the tied variables at will. All these alternatives are arbitrary, but they do permit the continuation of the solution by the simplex method. However, no matter which of the tied variables is removed, we shall encounter another difficulty in the next tableau. What will happen is that one of the remaining variables in the next tableau (the one corresponding to the tied variable that was not removed) will be reduced to a

---

* If we had denoted X, Y, Z by X_1, X_2, X_3, and S_1, S_2, S_3 by X_4, X_5, X_6, then S_1 would be removed first (Tableau 9.Ia), since it would have the smaller subscript of the tied variables.

magnitude of zero. Thus, when the key column in the next tableau is
chosen, we shall note that we really cannot introduce the new incoming
product, as the minimum nonnegative replacement ratio will be zero.
The procedure to follow at that point is to neglect this fact and proceed
to the third tableau with the assumption that the variable having zero
magnitude in the program (here S_1 in Tableau 9.II) has a very small
magnitude epsilon (ϵ), which can later be assumed to approach zero.   In
actual calculations, however, this ϵ need not appear.   This procedure will
become clear to the reader as we proceed with the actual solution of the
problem.

Since we have a tie between two variables only, let us proceed to solve
this problem from two directions, viz., by starting in one case with the
removal of S_3, and in the second case with the removal of S_1.

Case 1   Solution by Removing S_3 First

Various solution stages involved in removing S_3 first from Tableau 9.Ia are
given in Tableaux 9.Ib to 9.IV where we arbitrarily choose S_3 as the outgo-

Tableau 9.Ib

Program	Profit per unit	Quantity	22 X	30 Y	25 Z	0 S_1	0 S_2	0 S_3	
S_1	0	100	2	2	0	1	0	0	$\frac{100}{2}=50$
S_2	0	100	2	1	1	0	1	0	$\frac{100}{1}=100$
S_3	0	100	1	2	2	0	0	1	$\frac{100}{2}=50$
	Net-evaluation row:		22	30	25	0	0	0	

Outgoing variable   Incoming variable

ing variable.   In Tableau 9.II, since the entry in the net-evaluation row
under column X has the highest positive value (maximization problem),
variable X is the incoming variable.   Further, as the replacement ratios
show, row S_1 is the limiting row.   We observe, however, that the replacement quantity is limited to zero; i.e., no units of X can actually be
"brought in."   Nevertheless, we proceed to Tableau 9.III and then to

Tableau 9.IV by following the transformation rules discussed in Chapter 6.*

The value of the objective function for Tableau 9.III is the same as that for Tableau 9.II, namely, 1,500.   However, the basis of Tableau 9.III is

Tableau 9.II

Program	Profit per unit	Quantity	22 X	30 Y	25 Z	0 S_1	0 S_2	0 S_3	
S_1	0	0	1	0	-2	1	0	-1	$\frac{0}{1}=0\checkmark$
S_2	0	50	$\frac{3}{2}$	0	$0-1$	0	1	$-\frac{1}{2}$	$\frac{50}{\frac{3}{2}}=\frac{100}{3}$
Y	30	50	$\frac{1}{2}$	1	1	0	0	$\frac{1}{2}$	$\frac{50}{\frac{1}{2}}=100$
Net-evaluation row:			7	0	-5	0	0	-15	

Outgoing product   Incoming product

Tableau 9.III

Program	Profit per unit	Quantity	22 X	30 Y	25 Z	0 S_1	0 S_2	0 S_3	
X	22	0	1	0	-2	1	0	-1	
S_2	0	50	0	0	3	$-\frac{3}{2}$	1	1	$\frac{50}{3}=16\frac{2}{3}\checkmark$
Y	30	50	0	1	2	$-\frac{1}{2}$	0	1	$\frac{50}{2}=25$
Net-evaluation row:			0	0	9	-7	0	-8	

Outgoing variable   Incoming variable

different from the basis of Tableau 9.II.   In degenerate problems we may have to go through *several* such iterations, during which, although a new basis is acquired, the value of the objective function stays constant. Fortunately, as we see in Tableau 9.IV, such is not the case in our prob-

* In other words, row S_1 of Tableau 9.II is considered as the key row, and rows S_2 and Y are treated as nonkey rows. The rules of transformation are then applied to derive Tableau 9.III.

Tableau 9.IV

Program	Profit per unit	Quantity	22 X	30 Y	25 Z	0 S_1	0 S_2	0 S_3
X	22	$\frac{100}{3}$	1	0	0	0	$\frac{2}{3}$	$-\frac{1}{3}$
Z	25	$\frac{50}{3}$	0	0	1	$-\frac{1}{2}$	$\frac{1}{3}$	$\frac{1}{3}$
Y	30	$\frac{50}{3}$	0	1	0	$\frac{1}{2}$	$-\frac{2}{3}$	$\frac{1}{3}$

Net-evaluation row: 0   0   0 $-\frac{5}{2}$ -3 -11

lem.   Since all the entries in the net-evaluation row of Tableau 9.IV are nonpositive (maximization case), the optimal solution has been obtained.

Case 2   Solution by Removing S_1 First

The various solution stages involved in removing S_1 first from Tableau 9.Ia are given in Tableaux 9.Ic and 9.V to 9.VII.   In Tableau 9.Ic we arbi-

Tableau 9.Ic

Program	Profit per unit	Quantity	22 X	30 Y	25 Z	0 S_1	0 S_2	0 S_3	
S_1	0	100	2	2	0	1	0	0	$\frac{100}{2} = 50$
S_2	0	100	2	1	1	0	1	0	$\frac{100}{1} = 100$
S_3	0	100	1	2	2	0	0	1	$\frac{100}{2} = 50$

Net-evaluation row: 22   30   25   0   0   0

Outgoing variable   Incoming variable

trarily choose S_1 as the outgoing variable.   In Tableau 9.V, since the entry in the net-evaluation row under column Z has the largest positive value (maximization case), variable Z is the incoming variable.   Further, as the replacement ratios show, row S_3 is the limiting row.   We observe, however, that the replacement quantity of S_3 is limited to zero; i.e., no

Tableau 9.V

Program	Profit per unit	Quantity	22 X	30 Y	25 Z	0 S_1	0 S_2	0 S_3	
Y	30	50	1	1	0	$\frac{1}{2}$	0	0	
S_2	0	50	1	0	1	$-\frac{1}{2}$	1	0	$\frac{50}{1}=50$
S_3	0	0	-1	0	2	-1	0	1	$\frac{0}{2}=0\checkmark$
	Net-evaluation row:		-8	0	25	-15	0	0	

Outgoing variable   Incoming variable

Tableau 9.VI

Program	Profit per unit	Quantity	X	Y	Z	S_1	S_2	S_3	
Y	30	50	1	1	0	$\frac{1}{2}$	0	0	$\frac{50}{1}=50$
S_2	0	50	$\frac{3}{2}$	0	0	0	1	$-\frac{1}{2}$	$\frac{50}{\frac{3}{2}}=\frac{100}{3}\checkmark$
Z	25	0	$-\frac{1}{2}$	0	1	$-\frac{1}{2}$	0	$\frac{1}{2}$	
	Net-evaluation row:		$\frac{9}{2}$	0	0	$-\frac{5}{2}$	0	$-\frac{25}{2}$	

Outgoing variable   Incoming variable

Tableau 9.VII

Program	Profit per unit	Quantity	22 X	30 Y	25 Z	0 S_1	0 S_2	0 S_3
Y	30	$\frac{50}{3}$	0	1	0	$\frac{1}{2}$	$-\frac{2}{3}$	$\frac{1}{3}$
X	22	$\frac{100}{3}$	1	0	0	0	$\frac{2}{3}$	$-\frac{1}{3}$
Z	25	$\frac{50}{3}$	0	0	1	$-\frac{1}{2}$	$\frac{1}{3}$	$\frac{1}{3}$
	Net-evaluation row:	0	0	0	$-\frac{5}{2}$	-3	-11	

units of Z can actually be brought in.   Nevertheless, we proceed to Tableau 9.VI and then to Tableau 9.VII by following the rules of transformation discussed in Chapter 6.*

Since all the entries in the net-evaluation row are nonpositive (maximization case), an optimal solution has been obtained.

It should be noted that in this problem the resolution of degeneracy was a simple matter.   No matter which of the tied variables was removed first, we obtained the same solution in the *same number of iterations.* Two remarks must be made at this time.   First, an arbitrary selection of one of the tied variables may lead us to a longer path to the optimum solution.   In other words, arbitrary removal of one of the tied variables may mean that a much larger number of iterations will be necessary to arrive at the optimum solution than would be the case if some other tied variable were removed from the basis.   Second, a more serious situation may arise if an arbitrary selection of the tied variable leads us to what we earlier called cycling.   In cycling, we start from some basis and, after a few iterations, return to the same basis, so that an optimum solution may never be reached.

Although cycling is a theoretical possibility, it seldom occurs in practical problems.   However, general methods of resolving degeneracy have been devised which, if followed, will ensure against falling into the cycling process.   One such method was developed by Charnes and Cooper.† If the arrangement of the initial simplex tableau is exactly that given in this text (i.e., the identity matrix on the extreme right-hand side, and the main body adjacent to and left of the identity matrix), the procedure to be followed to resolve degeneracy is as follows:

1. Identify the tied variables or rows.
2. For each of the columns in the identity (starting with the extreme left-hand column of the identity and proceeding one at a time to the right), compute a ratio by dividing the entry in each tied row by the key-column number in that row.
3. Compare these ratios, column by column, proceeding to the right. The first time the ratios are unequal, the tie is broken.
4. Of the tied rows, the one in which the smaller algebraic ratio falls is the key row.
5. If the ratios in the identity do not break the tie, form similar ratios for the columns of the main body, and select the key row as described in steps 3 and 4.

* See previous footnote.
† A. Charnes et al., "An Introduction to Linear Programming," pp. 20–24, 62–69, John Wiley & Sons, Inc., New York, 1953.

The reader can verify that application of the above procedure for resolving degeneracy would mean that the variable S_3 should have been removed first when the tie developed in this example. We shall illustrate the procedure by applying it to Tableau 9.I*a*.

Step 1

Since the smallest nonnegative replacement ratio occurs in rows S_1 and S_3, we have a tie between these rows.

Step 2

For row S_1:

$$\tfrac{1}{2} = \tfrac{1}{2}$$

For row S_3:

$$\tfrac{0}{2} = 0$$

Steps 3 and 4

Since the smaller algebraic ratio in step 2 occurs for row S_3, row S_3 is the key row.

Once the key row has been identified by the application of the above rules, the tie is broken. We can then apply the simplex method in its regular form by following the rules of transformation discussed in Chapter 6. This was done in Tableaux 9.II through 9.IV.

Note that the tie in Tableau 9.I*a* was broken in step 4 of the procedure for resolving degeneracy. However, the tie between the rows may appear at any solution stage, and it may be necessary to apply all five steps before degeneracy is resolved.

chapter

10

The Transportation
Model

10.1 INTRODUCTION

The transportation model deals with a special class of linear-programming problems in which the objective is to "transport" a *single commodity* from various "origins" to different "destinations" at a minimum total cost.* The total supply available at the origins and the total quantity demanded by the destinations are given in the statement of the problem. Also given is the cost of shipping a unit of goods from a known origin to a known destination. As in the linear-programming problems discussed in previous chapters, all relationships are assumed to be linear.

With information about the total capacities of the origins, the total requirements of the destinations, and the shipping cost per unit of goods for available shipping routes, the transportation model is used to determine the optimum shipping program(s) resulting in minimum total shipping costs.

In so far as the transportation problem is a special case of the general linear-programming problem, it can always be solved by the simplex method. However, the *transportation algorithm*, which we shall develop in later sections of this chapter, provides a much more efficient method of handling such a problem. Let us now turn our attention to delineating the relationship between the general linear-programming problem and the transportation problem.

* Of course, if the payoff measure is of the "profit" variety, the objective will be to maximize total payoff.

160

10.2 TRANSPORTATION PROBLEM —A SPECIAL CASE

Having solved a general linear-programming problem by various methods in previous chapters, we observe again that such a problem always consists essentially of three components: (1) a linear objective function, (2) a set of linear structural constraints, and (3) a set of nonnegativity constraints. Let us illustrate these three components.

1   Linear Objective Function

Every linear-programming problem has as its objective the maximization or minimization of a linear objective function. This function is usually of the form

$$F(X) = \sum_{j=1}^{n} c_j x_j \qquad j = 1, 2, \ldots, n$$

where x_j = set of *structural variables;* these variables represent competing candidates or activities

c_j = set of so-called "price coefficients"; in the problem, c_j's are coefficients of structural variables in the objective function

A typical linear objective function involving n variables can be written as follows:

$$F(X) = c_1 x_1 + c_2 x_2 + \cdots + c_n x_n$$

2   Linear Structural Constraints

Every linear-programming problem contains a set of linear constraints. They embody the technical specifications and resource capacities of the problem structure and are therefore called *structural* constraints. These constraints are of the form

$$\sum_{j=1}^{n} \sum_{i=1}^{m} a_{ij} x_j \leq b_i \qquad \begin{cases} i = 1, 2, \ldots, m \\ j = 1, 2, \ldots, n \end{cases}$$

where the a_{ij}'s are a set of *structural coefficients* reflecting the technical specifications of the problem, and they appear as coefficients of the structural variables in the structural constraints.   The b_i's are a set of *constants* reflecting the maximum resource capacities or minimum resource requirements.

An expanded form of the linear structural constraints is given below:*

$$a_{11}x_1 + a_{12}x_2 + \cdots + a_{1n}x_n \leq b_1$$
$$a_{21}x_1 + a_{22}x_2 + \cdots + a_{2n}x_n \leq b_2$$
. . . .
$$a_{m1}x_1 + a_{m2}x_2 + \cdots + a_{mn}x_n \leq b_m$$

3   Nonnegativity Constraints

The structural variables, slack variables, and artificial slack variables of all linear-programming problems are restricted to nonnegative values. This is accomplished by imposing nonnegativity constraints of the form

$$x_j \geq 0 \qquad j = 1, 2, \ldots, n$$

If we let x_j's denote the structural variables, S_i's denote the slack variables, and A_i's denote the artificial slack variables, we may write these constraints as follows:

$$x_1 \geq 0 \qquad x_2 \geq 0 \qquad \cdots \qquad x_n \geq 0$$
$$S_1 \geq 0 \qquad S_2 \geq 0 \qquad \cdots \qquad S_m \geq 0$$
$$A_1 \geq 0 \qquad A_2 \geq 0 \qquad \cdots \qquad A_m \geq 0$$

Two remarks must be made at this time in reference to the general linear-programming problem.   First, the structural coefficients a_{ij} are not restricted to any particular value or values.   For example, a particular a_{ij} may be specified to have a value of 10, 20, 1, or 0.   Second, no restrictions are imposed regarding the homogeneity of units among the various inequalities representing the structural constraints.   Of the given constraints, some may refer to available capacities of machines performing different kinds of operations, while others may specify different types of, say, chemical characteristics.   In other words, the units of any one

---

* The structural constraints in their original form can, of course, be simple equalities or inequalities of the "less than or equal to" or "greater than or equal to" type.

constraint may not be the same as those of the other constraints and hence may not be interchangeable with the units of any other constraint. We can illustrate this by referring back to the vitamin problem of Chapter 7, in which the two structural constraints were concerned with different types of vitamins.

The transportation problem, in comparison with the general linear-programming problem, restricts the values that can be assigned to the structural coefficients and limits the constraints to only one type of units. In particular, the general linear-programming problem can be reduced to what is called a *transportation problem* if (1) the a_{ij}'s (coefficients of the structural variables in the constraints) are restricted to the values 0 and 1 and (2) there exists a homogeneity of units among the constraints.

The transportation problem and the general linear-programming problem can be compared by examining the simplex tableau (Figure 10.1) constructed from the transportation problem given in Table 10.1.

Let us now formulate a typical transportation problem involving three origins and four destinations.

10.3 THE PROBLEM

A manufacturing concern has three plants located in three different cities, all producing the same product. The total supply potential of the firm is absorbed by four large customers. Let us identify the three plants as O_1, O_2, and O_3, and the customers as D_1, D_2, D_3, and D_4. The relevant data on plant capacities, destination requirements, and shipping costs for individual shipping routes are recorded, in general terms, in Table 10.1.

As shown in the table, the matrix of our transportation problem has three rows and four columns and hence is *not* a square matrix. This emphasizes the point that in a transportation problem a given origin can simultaneously supply goods to more than one destination. As we shall see later, the so-called "assignment model" is restricted to a square matrix in the sense that one origin cannot simultaneously associate with more than one destination in the assignment problem.

Notice that the first subscript in each symbol used in Table 10.1 refers to the specific origin, and the second subscript to the particular destination. For example, c_{12} is the cost of shipping 1 unit of goods from origin O_1 to destination D_2, and the variable x_{34} is the quantity to be shipped from origin O_3 to destination D_4. Origin capacities and destination requirements are given along the outside (rims) of Table 10.1 and are usually referred to as *rim requirements*. Our problem is to choose that strategy (a particular program of shipping) which will satisfy the rim requirements at a minimum total cost.

Analysis of the Problem

The transportation problem given above, like the general linear-programming problem, consists of three components. First, we can formulate a linear objective function which is to be minimized. This function will represent the total shipping cost of all the goods to be sent from the origins to the destinations. Second, we can write a set of linear structural constraints. Of the seven constraints of this problem, three (one for

Table 10.1

Origin	Destination				Origin capacity per time period
	D_1	D_2	D_3	D_4	
O_1	c_{11} x_{11}	c_{12} x_{12}	c_{13} x_{13}	c_{14} x_{14}	b_1
O_2	c_{21} x_{21}	c_{22} x_{22}	c_{23} x_{23}	c_{24} x_{24}	b_2
O_3	c_{31} x_{31}	c_{32} x_{32}	c_{33} x_{33}	c_{34} x_{34}	b_3
Destination requirement per time period	d_1	d_2	d_3	d_4	

c_{ij} = cost of shipping a unit of goods from ith origin to jth destination
x_{ij} = number of units to be shipped from ith origin to jth destination

Assume

$$\Sigma b_i = \Sigma d_j$$

That is, total origin capacities equal total destination requirements, and $i = 1, 2, 3$; $j = 1, 2, 3, 4$.

each row) will give the relationships between the origin capacities and the goods to be received by different destinations. These are called *capacity* constraints. The other four constraints (one for each column) will specify the relationships between destination requirements and the goods to be shipped from different origins. These are called *requirement* constraints. Third, we can specify a set of nonnegativity constraints for the structural variables x_{ij}. They will state that no negative shipment is permitted. The general correspondence between a typical linear-programming problem and the transportation problem is thus complete.

The three component parts of our transportation problem are given below:

Minimize

$$F(X) = c_{11}x_{11} + c_{12}x_{12} + c_{13}x_{13} + c_{14}x_{14} + c_{21}x_{21} + c_{22}x_{22} + c_{23}x_{23}$$
$$+ c_{24}x_{24} + c_{31}x_{31} + c_{32}x_{32} + c_{33}x_{33} + c_{34}x_{34}$$

subject to

$$
\begin{array}{llll}
x_{11} + x_{12} + x_{13} + x_{14} & & = b_1 & (1) \\
\quad x_{21} + x_{22} + x_{23} + x_{24} & & = b_2 & (2) \\
\quad\quad x_{31} + x_{32} + x_{33} + x_{34} & = b_3 & (3) \\
x_{11} \quad\quad + x_{21} \quad\quad + x_{31} & = d_1 & (4) \\
\quad x_{12} \quad\quad + x_{22} \quad\quad + x_{32} & = d_2 & (5) \\
\quad\quad x_{13} \quad\quad + x_{23} \quad\quad + x_{33} & = d_3 & (6) \\
\quad\quad\quad x_{14} \quad\quad + x_{24} \quad\quad + x_{34} & = d_4 & (7)
\end{array}
$$

and $x_{ij} \geq 0$; $i = 1, 2, 3$; $j = 1, 2, 3, 4$.

10.4 BUILDING A SIMPLEX TABLEAU FOR THE TRANSPORTATION PROBLEM

Since Equations (1) through (3) refer to origin capacities, we can think of these as inequalities of the "less than or equal to" type asserting the fact that different origins cannot produce more than their respective capacities. For purposes of construction of the simplex tableau, therefore, we may modify Equations (1) through (3) with the addition of slack variables S_1, S_2, and S_3, respectively. The cost coefficients of these slack variables are, of course, zero. Variables S_1, S_2, and S_3 represent the respective idle capacities of origins O_1, O_2, and O_3.

On the other hand, we should consider Equations (4) through (7) as strict equations or inequalities of the "greater than or equal to" type. Long-run interests of the company require that it be willing to supply exactly that quantity of goods which is specified by each customer (or perhaps more, if circumstances demand). For purposes of construction of the first simplex tableau, therefore, Equations (4) through (7) may be modified with the addition of nonnegative artificial slack variables A_1, A_2, A_3, and A_4, respectively.* The cost coefficient of each of these artificial slack variables is obviously M.

Program	Cost per unit	Quantity	c_{11} x_{11}	c_{12} x_{12}	c_{13} x_{13}	c_{14} x_{14}	c_{21} x_{21}	c_{22} x_{22}	c_{23} x_{23}	c_{24} x_{24}	c_{31} x_{31}	c_{32} x_{32}	c_{33} x_{33}	c_{34} x_{34}	0 S_1	0 S_2	0 S_3	M A_1	M A_2	M A_3	M A_4
S_1	0	b_1	1	1	1	1	0	0	0	0	0	0	0	0	1	0	0	0	0	0	0
S_2	0	b_2	0	0	0	0	1	1	1	1	0	0	0	0	0	1	0	0	0	0	0
S_3	0	b_3	0	0	0	0	0	0	0	0	1	1	1	1	0	0	1	0	0	0	0
A_1	M	d_1	1	0	0	0	1	0	0	0	1	0	0	0	0	0	0	1	0	0	0
A_2	M	d_2	0	1	0	0	0	1	0	0	0	1	0	0	0	0	0	0	1	0	0
A_3	M	d_3	0	0	1	0	0	0	1	0	0	0	1	0	0	0	0	0	0	1	0
A_4	M	d_4	0	0	0	1	0	0	0	1	0	0	0	1	0	0	0	0	0	0	1

Figure 10.1

With this completed, the data of our problem can be reflected in the first simplex tableau, as shown in Figure 10.1. We note that all entries in the initial simplex tableau are either 1 or 0. The 1s appear in the form of scattered rows and slanted diagonals. Further, each column, except those in the identity part of the tableau, represents a column vector in which two elements are 1 and the rest are zero.† Whenever the initial simplex tableau appears in a form such as that of Figure 10.1, the linear-programming problem can be classified as a transportation problem.

The assignment of specific numerical values to origin capacities (b_1, b_2,

---

* We assume exact equality in the original formulations of the requirement constraints.
† This particular characteristic of the transportation problem forms the basis of the so-called "modified-distribution method" of solving such problems (see Section 10.8).

b_3), destination requirements (d_1, d_2, d_3, d_4), and cost coefficients (c_{ij}'s) in Table 10.1 would give us a concrete transportation problem whose *initial* solution by the simplex method would be given by Figure 10.1. Further iterations following the rules of the simplex method would, no doubt, yield the optimum solution to this problem. This would be a rather lengthy process and would not add anything new to our knowledge of the simplex method. Fortunately, however, a simple and routine method of solving such problems has been developed. It is fittingly called the *transportation model*. Whenever a given linear-programming problem can be placed in the transportation framework, it is far simpler to solve it by the transportation method than by the simplex method. Before we describe and develop the transportation method, let us comment on certain characteristics of the transportation problem and its solution.

First, a little reflection will show that for the transportation problem of Table 10.1 only six rather than seven structural constraints need be specified. In view of the fact that the sum of the origin capacities equals the sum of the destination requirements ($\Sigma b_i = \Sigma d_j$), any solution satisfying six of the seven constraints will automatically satisfy the last constraint. In general, therefore, if m represents the number of rows and n represents the number of columns in a given transportation problem, we can state the problem completely with $m + n - 1$ equations. This means that one of the rows of the simplex tableau in Figure 10.1 represents a redundant constraint and, hence, can be deleted. This also means that a basic feasible solution of a transportation problem has only $m + n - 1$ positive components.

Second, if origin capacities equal destination requirements, it is always possible to design an initial basic feasible solution in such a manner that the rim requirements are satisfied. This can be accomplished either by inspection or by following certain formal methods for making the initial allocation. Three such methods, the so-called "northwest-corner" rule, the "Vogel's approximation method," and the "inspection" method, will be described later.

10.5 APPROACH OF THE TRANSPORTATION METHOD

The transportation method consists of three basic steps. The first step involves making the initial shipping assignment in such a manner that a basic feasible solution is obtained. This means that $m + n - 1$ cells (routes) of the transportation matrix are used for shipping purposes.

The cells having the shipping assignment will be called *occupied* cells, while the remaining cells of the transportation matrix will be referred to as *empty* cells.

The purpose of the second step is to determine the opportunity costs associated with the empty cells. The opportunity costs of the empty cells can be calculated individually for each cell or simultaneously for the whole matrix. If the opportunity costs of all the empty cells are nonpositive, we can be confident that an optimum solution has been obtained.* On the other hand, if even a single empty cell has a positive opportunity cost, we proceed to step 3.

The third step involves determining a new and better basic feasible solution. Once this new basic feasible solution has been obtained, we repeat steps 2 and 3 until an optimum solution has been designed.

The remaining sections of this chapter are devoted to illustrating the development and application of the above-mentioned approach to the solution of a given transportation problem.

10.6 METHODS OF MAKING THE INITIAL ASSIGNMENT

The first step in the transportation method, as stated above, consists in making an initial assignment in such a manner that a basic feasible solution (number of occupied cells equals $m + n - 1$) is obtained. Various methods of making such an assignment are available. We shall discuss three such methods in considering the transportation problem in Table 10.2.

Northwest-corner Rule

According to this rule, the first allocation is made to the cell occupying the upper left-hand (northwest) corner of the matrix. Further, this allocation is of such a magnitude that either the origin capacity of the first row is exhausted or the destination requirement of the first column is satisfied or both. If the origin capacity of row 1 is exhausted first, we move down the first column and make another allocation which either exhausts the origin capacity of row 2 or satisfies the remaining destination

---

* The transportation problem falls under the category of decision making under *certainty*. Hence, the optimum solution must not be associated with positive opportunity costs.

requirement of column 1. On the other hand, if the first allocation completely satisfies the destination requirement of column 1, we move to the right in row 1 and make a second allocation which either exhausts the remaining capacity of row 1 or satisfies the destination requirement of column 2, and so on. In this manner, starting from the upper left-hand corner of the given transportation matrix, satisfying the individual destination requirements, and exhausting the origin capacities *one at a time,*

Table 10.2

Origin	Destination					Origin capacity per time period
	D_1	D_2	D_3	D_4	D_5	
O_1	12 40	4 15	9	5	9	55
O_2	8 5	1	6 40	6	7	45
O_3	1	12	4 6	7 20	7	30
O_4	10	15	6	9 10	1 40	50
Destination requirement per time period	40	20	50	30	40	

we move toward the lower right-hand corner until all the rim requirements are satisfied. It should be noted that when we follow the northwest-corner rule we pay no attention to the relative costs of the different routes while making the first assignment.

For the transportation problem of Table 10.2, application of the north-west-corner rule dictates that we first "load" or "fill" cell O_1D_1, which lies in the upper left-hand (northwest) corner. The product requirement of D_1 is 40 units, and the capacity of O_1 is 55 units; the lower of these two numbers, that is, 40, is placed in cell O_1D_1. This means that the requirement of D_1 is fully satisfied, but we still have 15 units (55 − 40) of unused capacity at O_1. Thus, we move to the right of cell O_1D_1 in the first row.

At this stage we note that the destination requirement of column D_2 is 20 units.   Knowing that 15 units of capacity O_1 are still unused, we route all 15 units to destination D_2 (place 15 in cell O_1D_2).   This completely exhausts the capacity O_1, but column D_2 still needs 5 units (20 − 15) to satisfy its requirement.   Thus, we move down column D_2 and supply these 5 units from capacity O_2 (place 5 in cell O_2D_2).   This leaves 40 units of unused capacity at Q_2; these are routed to D_3 (place 40 in cell

Table 10.3   *Initial Assignment by Northwest-corner Rule*

Origin	Destination					Total
	D_1	D_2	D_3	D_4	D_5	
O_1	12 ⓐ40	4 ⓐ15	9	5	9	55
O_2	8	1 ⓐ5	6 ⓐ40	6	7	45
O_3	1	12 ⓐ10	4 ⓐ20	7	7	30
O_4	10	15	6 ⓐ10	9 ⓐ40	1	50
Total	40	20	50	30	40	180 / 180

O_2D_3).   The remaining requirement of 10 units (50 − 40) for D_3 is supplied from O_3 (place 10 in cell O_3D_3).   This leaves 20 units of unused capacity at O_3; these are routed to D_4 (place 20 in cell O_3D_4).   The remaining requirement of 10 units (30 − 20) for D_4 is supplied from O_4 (place 10 in cell O_4D_4).   We are now left with 40 units of unused capacity at O_4; these are finally routed to D_5 (place 40 in cell O_4D_5).   The entire table has now been loaded, resulting in the initial program given in Table 10.3.   The circled numbers in the table give the number of units shipped from a particular origin to a certain destination.   The cells in which these circled numbers are entered are our occupied cells.   The rest of the cells are the empty cells.

It is to be observed that the number of occupied cells is

$$m + n - 1 = 4 + 5 - 1 = 8$$

$$= (\text{number of rows} + \text{number of columns} - 1)$$

The solution at this stage is therefore not degenerate.*

The total cost of this assignment is

$$1{,}095 =$$

$$40 \times 12 + 15 \times 4 + 5 \times 1 + 40 \times 6 + 10 \times 4 + 20 \times 7 + 10 \times 9 + 40 \times 1$$

It should be noted that the last allocation (cell $O_4 D_5$) *simultaneously* satisfied the requirement of column D_5 and exhausted the capacity of O_4. This is the "normal" situation in the last allocation made by the north-west-corner rule; if this occurs *only* in the last allocation, we can be certain of having a basic feasible solution. However, if any allocation previous to the last allocation happens to be such that it simultaneously satisfies the requirement of some destination and exhausts the capacity of some origin, then the number of occupied cells will be less than $m + n - 1$. This will mean that we have a degenerate basic feasible solution.† The reader should try to make an initial assignment in Table 10.2 by following the northwest-corner rule after having changed the destination require-ments of D_2 and D_3 to 15 and 55 units, respectively.

Vogel's Approximation Method (VAM)

According to this method, a *difference column* and a *difference row* repre-senting the difference between the costs of the *two cheapest routes* for each origin and destination are computed. Each individual difference can be thought of as a penalty‡ for not using the cheapest route. After all such penalty ratings have been computed for the given data, the highest difference or penalty rating is identified. Then the lowest-cost cell in

---

* A basic feasible solution for a transportation problem requires only $m + n - 1$ positive components. Thus, whenever a transportation program has $m + n - 1$ occupied cells, the solution is not degenerate.

† A method for resolving degeneracy in transportation problems is discussed later in this chapter (see Section 10.13).

‡ An-Min Chung, "Linear Programming," p. 248, Charles E. Merrill Books, Inc., Columbus, Ohio, 1963.

the second cheapest — the cheapest one

Table 10.4a

	D_1	D_2	D_3	D_4	D_5	Difference column or penalty
O_1	[12]	[4]	[9]	[5]	[9]	1
O_2	[8]	[1]	[6]	[6]	[7]	5
O_3	([1])	[12]	[4]	[7]	[7]	3
O_4	[10]	[15]	[6]	[9]	[1]	5
Difference row or penalty	7 ✓	3	2	1	6	

	D_1	D_2	D_3	D_4	D_5	Capacity
O_1	[12]	[4]	[9]	[5]	[9]	55
O_2	[8]	[1]	[6]	[6]	[7]	45
O_3	[1] (30)	[12]	[4]	[7]	[7]	30 0
O_4	[10]	[15]	[6]	[9]	[1]	50
Requirement	40 10	20	50	30	40	

Table 10.4b

	D_1	D_2	D_3	D_4	D_5	Difference column or penalty
O_1	[12]	[4]	[9]	[5]	[9]	1
O_2	[8]	[1]	[6]	[6]	[7]	5
O_4	[10]	[15]	[6]	[9]	([1])	5
Difference row or penalty	2	3	0	1	6 ✓	

	D_1	D_2	D_3	D_4	D_5	Capacity
O_1	[12]	[4]	[9]	[5]	[9]	55
O_2	[8]	[1]	[6]	[6]	[7]	45
O_4	[10]	[15]	[6]	[9]	[1] (40)	50 10
Requirement	10	20	50	30	40 0	

Table 10.4c

	D_1	D_2	D_3	D_4	Difference column or penalty
O_1	[12]	[4]	[9]	[5]	1
O_2	[8]	[1]	[6]	[6]	5 ✓
O_4	[10]	[15]	[6]	[9]	3
Difference row or penalty	2	3	0	1	

	D_1	D_2	D_3	D_4	Capacity
O_1	[12]	[4]	[9]	[5]	55
O_2	[8]	[1] (20)	[6]	[6]	45 25
O_4	[10]	[15]	[6]	[9]	10
Requirement	10	20 0	50	30	

that row or column in which the highest penalty rating* was placed is the cell to which the first assignment is made. This assignment either exhausts the capacity of an origin or meets the requirement of a destina-

---

* Should there be a tie for highest penalty rating or difference value, we can arbitrarily choose one to break the tie. Although rules for breaking ties are available, it is usually easier simply to pick one of the tied columns or rows for making the allocation. See N. V. Reinfeld and William R. Vogel, "Mathematical Programming," chap. 4, Prentice-Hall, Inc., Englewood Cliffs, N.J., 1958.

Table 10.4*d*

	D_1	D_3	D_4	Difference column or penalty
O_1	12	9	5	4 ✓
O_2	8	6	6	0
O_4	10	6	9	3
Difference row or penalty	2	0	1	

	D_1	D_3	D_4	Capacity
O_1	12	9	5 ㉚	5̶5̶ 25
O_2	8	6	6	25
O_4	10	6	9	10
Requirement	10	50	3̶0̶ 0	

Table 10.4*e*

	D_1	D_3	Difference column or penalty
O_1	12	9	3
O_2	8	6	2
O_4	10	6	4 ✓
Difference row or penalty	2	0	

	D_1	D_3	Capacity
O_1	12	9	25
O_2	8	6	25
O_4	10	6 ⑩	1̶0̶ 0
Requirement	10	5̶0̶ 40	

Table 10.4*f*

	D_1	D_3	Difference column or penalty
O_1	12	9	3
O_2	8	6	2
Difference row or penalty	4 ✓	3	

	D_1	D_3	Capacity
O_1	12	9	25
O_2	8 ⑩	6	2̶5̶ 15
Requirement	1̶0̶ 0	40	

Table 10.4*g*

	D_3	Capacity
O_1	9 ㉕	25
O_2	6 ⑮	15
Requirement	40	

tion or both.   The particular row or column which has been thus satisfied is removed from the transportation matrix.   The process is then repeated until an initial assignment using $m + n - 1$ routes has been obtained. This approach has the disadvantage of necessitating some computational work before the initial program is obtained, but it usually results in the attainment of the optimal program in fewer iterations than are required when the initial program is obtained by using the northwest-corner rule.

Table 10.5   *Initial Assignment by VAM*

Origin	Destination					Total
	D_1	D_2	D_3	D_4	D_5	
O_1	12	4	9 (25)	5 (30)	9	55
O_2	8 (10)	1 (20)	6 (15)	6	7	45
O_3	1 (30)	12	4	7	7	30
O_4	10	15	6 (10)	9 (40)	1	50
Total	40	20	50	30	40	180 / 180

The mechanics for obtaining the initial assignment for the transportation problem of Table 10.2 by Vogel's approximation method (VAM) is illustrated in Tables 10.4a through 10.4g.   In Table 10.4a, the highest difference or penalty rating is 7, and this falls under column D_1.   The first allocation, therefore, must be made to that cell in column D_1 which has the lowest shipping cost.   Since cell O_3D_1 has the lowest shipping cost in that column, we now compare the capacity of O_3 (30 units) with the requirement of D_1 (40 units).   The lower of the two numbers, that is, 30, is placed in cell O_3D_1.   This means that the capacity of O_3 has been fully utilized, and row O_3 can be removed temporarily from the transportation matrix.   Column D_1, however, cannot be removed, since we still need 10 units to satisfy its requirements fully.

We now have arrived at Table 10.4b. The process of computing penalties is repeated, and in Table 10.4b we note that the highest penalty falls under column D_5. We therefore make an assignment in the lowest-cost cell of column D_5. This assignment (place 40 in cell O_4D_5) is such that column D_5 can now be removed from the matrix, and we proceed to Table 10.4c. By repeating the same process in Tables 10.4c to 10.4g, we finally obtain the assignment in Table 10.5. Observe that the number of occupied cells is $m + n - 1 = 4 + 5 - 1 = 8$. The initial solution is therefore a basic feasible solution, and the problem at this stage is not degenerate. The total cost of this assignment is \$695—considerably less than the total cost associated with the initial program obtained via the northwest-corner rule.

Initial Assignment by Inspection

One can, no doubt, make an initial assignment in a transportation problem simply by inspection and judgment. This is, needless to say, not a formal method of obtaining an initial assignment. However, for transportation problems of small dimensions, it has the advantage of speed. The first allocation is made to that cell whose shipping cost per unit is lowest. This lowest-cost cell is loaded or filled as much as possible in view of the origin capacity of its row and the destination requirement of its column. Then we move to the next lowest-cost cell and make an allocation in view of the remaining capacity and requirement of its row and column, and so on. Should there be a tie for lowest-cost cell during any allocation, we can exercise "judgment" in breaking the tie or we can arbitrarily choose a cell for allocation. The total number of allocations, of course, must be such that a basic feasible solution ($m + n - 1$ occupied cells) is obtained. Let us illustrate the inspection method for the transportation problem of Table 10.2. We note that cells O_2D_2, O_3D_1, and O_4D_5 each have a shipping cost of \$1 per unit. Thus there is a tie for the first allocation. We arbitrarily choose cell O_3D_1 for the first allocation and route 30 units from O_3 to D_1.* This means that the capacity of O_3 is fully utilized (cross off row O_3 with a light pencil). For the second allocation, we observe that there is a tie between cells O_2D_2 and O_4D_5. We arbitrarily choose O_4D_5 and ship 40 units through this route. This completely satisfies the requirement of column D_5 (cross off column D_5 with a light pencil). For the third allocation, we note that cell O_2D_2 is

---

* In this case, one could easily have exercised judgment in terms of the VAM penalty ratings associated with the tied cells. As a matter of fact, one should exercise such a judgment in all stages of the inspection method.

now the lowest-cost cell and, therefore, we ship 20 units through this route (place 20 in cell O_2D_2 and cross off column D_2). Of those remaining, cell O_1D_4 has the lowest cost, and we route 30 units through O_1D_4 (place 30 in cell O_1D_4 and cross off column D_4). Next, we observe that there is a tie between O_2D_3 and O_4D_3 for the fifth allocation. We arbitrarily choose cell O_4D_3 and ship 10 units (the remaining capacity of O_4) through this route (place 10 in cell O_4D_3 and cross off row O_4). Of those remain-

Table 10.6   *Initial Assignment by Inspection*

Origin	Destination					Total
	D_1	D_2	D_3	D_4	D_5	
O_1	12 (10)	4	9 (15)	5 (30)	9	5̶5̶ 2̶5̶ 1̶0̶ 0
O_2	8	1 (20)	6 (25)	6	7	4̶5̶ 2̶5̶ 0 ······6th
O_3	1 (30)	12	4	7	7	3̶0̶ 0 ······1st
O_4	10	15	6 (10)	9	1 (40)	5̶0̶ 1̶0̶ 0 ······5th
Total	4̶0̶ 1̶0̶ 0	2̶0̶ 0	5̶0̶ 4̶0̶ 1̶5̶ 0	3̶0̶ 0	4̶0̶ 0	
	8th	3d	7th	4th	2d	

ing, cell O_2D_3 has the lowest cost, and we route 25 units (the remaining capacity of row O_2) through cell O_2D_3 (place 25 in cell O_2D_3 and cross off row O_2). We are now left with 25 units at O_1, while D_1 and D_3 still require 10 and 15 units, respectively. Hence, we route 10 units through O_1D_1 and 15 units through O_1D_3. All the rim requirements have now been satisfied, and we have the initial assignment in Table 10.6. The dotted lines crossing the cost squares (c_{ij}'s) have been numbered to show the order in which different rows and columns were crossed off, as an aid to making the initial assignment by inspection.

It is to be observed that the number of occupied cells is 8 (that is,

$m + n - 1$), and thus we have a basic feasible solution. The total cost of this assignment is $705. It should be compared with the total costs associated with the initial solutions obtained by the northwest-corner rule ($1,095) and Vogel's approximation method ($695).

It will be recalled that the approach of the transportation method is based on three steps: (1) making the initial assignment in order to obtain a basic feasible solution, (2) determining the opportunity costs of the empty cells, and (3) designing a better basic feasible solution (provided step 2 indicates that the program can be improved) and repeating steps 2 and 3 until an optimal solution has been obtained. The application of the first step has now been illustrated in connection with the transportation problem of Table 10.2. Next we shall show the application of steps 2 and 3 to complete our illustration of the transportation method. There are, however, two methods of carrying out steps 2 and 3. One is called the *steppingstone* method, whereas the other is referred to as the *modified-distribution* method. We shall first discuss the steppingstone method.

10.7 STEPPINGSTONE METHOD FOR OBTAINING AN OPTIMAL SOLUTION

To illustrate the steppingstone method, we shall first solve the very simple transportation problem given in Table 10.7. The method will then be used in deriving the optimum solution to our problem of Table 10.2. The purpose of solving the simple problem of Table 10.7 is to

Table 10.7

Origin	Destination		Origin capacity per time period
	D_1	D_2	
O_1	2	2	1,000
O_2	1	2	600
Destination requirement per time period	900	700	1,600 1,600

familiarize the reader with the terminology and rationale of the stepping-stone method.

Following the northwest-corner rule, we obtain the initial program in Table 10.8.   The circled numbers within the body of the matrix refer to the specific allocations of the first program.   This program calls for shipping 900 units from O_1 to D_1, 100 units from O_1 to D_2, and 600 units from O_2 to D_2.   Obviously, this program satisfies all the rim requirements. Note further that the number of occupied cells is 3, which is 1 less than

Table 10.8

Origin	Destination		Total
	D_1	D_2	
O_1	2 ⟨900⟩	2 ⟨100⟩	1,000
O_2	1	2 ⟨600⟩	600
Total	900	700	

the sum of the numbers of rows and columns.   In other words, the number of occupied cells in this program equals $m + n - 1$.*   Thus we have a basic feasible solution.

* As we have established previously, only $m + n - 1$ equations are needed to state a transportation problem.   The present problem can be stated with the following equations:

$$x_{11} + x_{12} = 1,000$$

$$x_{21} + x_{22} = 600$$

$$x_{11} + x_{21} = 900$$

Naturally, letting $x_{21} = 0$ gives a solution in which $x_{11} = 900$, $x_{22} = 600$, and $x_{12} = 100$.

Determining the Opportunity Cost of the Empty Cells

Is the above program an optimal program? To answer this question, we must apply step 2, that is, determine the opportunity costs of the empty cells. In so far as the transportation model involves decision making under certainty, we know that an optimal solution must not incur any positive opportunity cost. Thus, to determine whether any positive opportunity cost is associated with a given program, we must test the empty cells (cells representing routes not used in the given program) of the transportation matrix for the presence or absence of opportunity cost. The absence of positive opportunity costs in *all empty cells* will indicate that an optimal solution has been obtained. If, on the other

Table 10.9

Take 1 unit out of O_2D_2: -1
Add 1 unit to O_2D_1: $+1$
Take 1 unit out of O_1D_1: -1
Add 1 unit to O_1D_2: $+1$

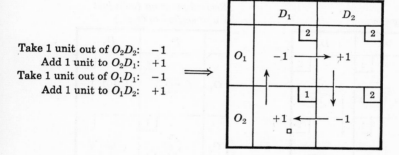

hand, even a single empty cell has a positive opportunity cost, the given program is not the optimal program and, hence, should be revised.*

Let us examine our first program in view of the above discussion. Since cell O_2D_1 in this program is empty, we wish to determine whether or not there is an opportunity cost associated with it. This is accomplished by shifting 1 unit of goods to cell O_2D_1, making other shifts necessary to satisfy the rim requirements, and then finding the cost consequence of these changes. Let us shift 1 unit from cell O_2D_2 to cell O_2D_1. This shift will necessitate the changes noted in Table 10.9 in order to keep the rim requirements satisfied. These changes are associated with the following cost consequence or cost change:

$$-2 + 1 - 2 + 2 = -1 \text{ dollar}$$

* It will be recalled that the test for optimality in the simplex method was also based on the concept of opportunity cost.

Since the shifting of 1 unit to O_2D_1 yields a negative cost change, it is obviously a desirable shift. The fact that the transfer of 1 unit to cell O_2D_1 resulted in a net cost change of -1 dollar indicates that the opportunity cost of *not* including cell O_2D_1 in the first program is $+1$ dollar per unit of shipment. The empty cell O_2D_1 must, therefore, be included in a new and improved program.

Revision of the Given Program

Having discovered that the opportunity cost of the empty cell O_2D_1 is positive, we must next obtain a new basic feasible solution. This is done

Table 10.10

First program

Revised program (*with just 1 unit shifted to* O_2D_1)

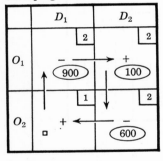

by designing a new improved program in which cell O_2D_1 is included in the shipping strategy. Let us make the improvement by shifting just 1 unit from cell O_2D_2 to cell O_2D_1. The revised program is given in Table 10.10. The shift of 1 unit from cell O_2D_2 to cell O_2D_1 means that we are left with 599 units in cell O_2D_2. This change, it should be noted, has not violated the capacity constraints of either row 1 or row 2. But what about the requirement constraints of columns 1 and 2? With the above change, we now have 1 unit in O_2D_1, 900 units in O_1D_1, 599 units in O_2D_2, and 100 units in O_1D_2. In other words, column 1 has 901 units (one more unit than the requirement of column 1), and column 2 has 699 units (one less unit than the requirement of column 2). Clearly, this situation can be remedied by shifting 1 unit from cell O_1D_1 to cell

O_1D_2, a change which will simultaneously satisfy the row and column requirements.

The revised program, with 1 unit shifted to cell O_2D_1, is shown in Table 10.10. The change in the program effected by the introduction of 1 unit to cell O_2D_1, as we established earlier, reduces the total shipping cost by $1. In so far as we gain this advantage each time a unit is shifted to cell O_2D_1, we must shift to cell O_2D_1 as many units as possible. As the *closed loop* (plus and minus signs connected by arrows) of Table 10.10 shows, we cannot shift more than 600 units to O_2D_1, for the allocation of more than 600 units to cell O_2D_1 would certainly violate the capacity constraint of row O_2.

Table 10.11

	D_1	D_2	Total
	2	2	
O_1	300	700	1,000
	1	2	
O_2	600		600
Total	900	700	

Our second program (a better basic feasible solution), the result of the above discussion, appears in Table 10.11. Is this the optimal allocation? The answer to this question can be obtained by testing the opportunity cost of cell O_2D_2, which is now the only empty cell. The answer is in the affirmative, since the opportunity cost of cell O_2D_2 is not positive. This can be verified by shifting 1 unit to cell O_2D_2 and noting that the net cost consequence of such a shift is +1 dollar ($+2 - 2 + 2 - 1$). The opportunity cost, being the negative of the corresponding net cost change, is therefore negative. Hence, the assignment of Table 10.11 gives an optimal solution with a total shipping cost of $2,600. No other program for this problem can result in a lower total shipping cost.

Let us recapitulate briefly the method of attack followed in solving this transportation problem. First, we designed a basic feasible solution by following the northwest-corner rule for making an initial assignment.*

Second, having obtained a basic feasible solution, we proceeded to determine the opportunity cost of the empty cell in order to determine whether the first program was an optimal program. The method employed to determine the opportunity cost of the empty cell consisted in (1) drawing a closed loop which passed through the empty cell and the adjacent occupied cells with proper plus and minus signs at the corners of the loop,† (2) shifting 1 unit of goods to the empty cell (accomplished by the addition of 1 unit to all those cells in which fell a plus sign of the closed loop and by the subtraction of 1 unit from all cells in which fell a minus sign), (3) determining the net cost change associated with shifting 1 unit to the empty cell, and (4) taking the negative of the net cost change in (3) to find the opportunity cost of the empty cell. In the simple transportation problem of Table 10.7, we had to find the opportunity cost of only one empty cell. In a problem of larger dimensions, the opportunity costs of all the empty cells must be determined by this procedure. The point to emphasize here is that a separate closed loop must be established for every empty cell (in the steppingstone method) before the opportunity costs of the empty cells can be determined.

Third, after ascertaining that the opportunity cost of the empty cell was positive, we changed the initial program by filling the empty cell

---

* In most cases, the initial assignment made by following the northwest-corner rule will be such that the number of occupied or filled cells equals $m + n - 1$, where m is the number of rows and n is the number of columns. When this happens, we have a basic feasible solution. If the number of occupied cells in the initial assignment is less than $m + n - 1$, the problem is said to be degenerate at the very beginning. This type of degeneracy, as well as the degeneracy occurring during the solution stages, can easily be resolved by judicious placement of a small number epsilon in the empty cell(s). The epsilon is to be placed in that cell(s) which will help complete the different loops for all the empty cells. An illustrative example is given in a later section (see Section 10.13).

† While tracing this closed loop, one should start with the empty cell being evaluated and draw an arrow from that empty cell to an occupied cell in the same row or column. Then, a plus sign is placed in the empty cell, and a negative sign in the occupied cell to which the arrow was drawn. Next, one moves horizontally or vertically (never diagonally) to another occupied cell, and so on, until one is back to the original empty cell. At each turn of the loop, plus and minus signs are placed alternately. Further, there is the important restriction that there are *exactly one positive* terminal and *exactly one negative* terminal in any row or column through which the loop happens to pass. Obviously, this restriction is imposed to ensure that the rim requirement will not be violated when the units are shifted (to obtain a new program) along this closed loop. Mechanically, this implies that during the tracing of the closed loop right-angle turns must be made *only* at the occupied cells. The starting point of a closed loop is identified by the symbol □ in this book.

(O_2D_1) as much as possible in view of the rim requirements.* The revision of the given program was guided by the plus and minus signs of the closed loop. The smallest of the numbers in the cells in which minus signs of the closed loop appeared (600) gave the total number of units to be shifted to the empty cell. The shifting was accomplished by adding this number (600) to all the cells containing the plus signs of the loop and subtracting it from all the cells containing the minus signs of the loop. These changes gave us our new basic feasible solution.

Finally, we tested the empty cell (O_2D_2) of the second program and found that its opportunity cost was not positive. We therefore came to the conclusion that an optimal solution to our problem had been obtained.

The procedure described above forms the core of the steppingstone method. Although the transportation problem that we solved was represented by only a 2×2 matrix (Table 10.7), the steppingstone method may be applied to any $m \times n$ matrix.

Suppose that our initial assignment in a 4×5 (4 rows and 5 columns) transportation problem results in 8 occupied cells and 12 empty cells. In order to test the optimality of this program and, then, to revise it, we must calculate the opportunity cost of each of the 12 empty cells. If we discover that the initial program can be improved, we revise the program by including that empty cell whose opportunity cost is highest. Note that, regardless of the number of empty cells having positive opportunity costs, only one cell at a time is included in the new program.

We shall now apply the steppingstone method to the problem of Table 10.2.

Step 1   Obtain an Initial Basic Feasible Solution

An initial basic feasible solution for a given transportation problem may be obtained by following the northwest-corner rule, by the application of Vogel's approximation method, or simply by inspection. It will be recalled that for the transportation problem of Table 10.2 we obtained three different basic feasible solutions, given in Tables 10.3, 10.5, and 10.6. Of these, let us take the basic feasible solution of Table 10.6 (initial assignment by inspection) as the starting point for obtaining the

---

* The fact that the opportunity cost of even a single cell is positive indicates that an optimum solution has not been obtained and that the given program must be revised to obtain a better basic feasible solution. Normally, the improved program will include that empty cell whose opportunity cost is highest. Since in this problem we had only one empty cell O_2D_1, the new program included that cell.

Table 10.12 *First Program*

Origin	Destination					Total
	D_1	D_2	D_3	D_4	D_5	
O_1	12 ⑩	4	9 ⑮	5 ㉚	9	55
O_2	8	1 ⑳	6 ㉕	6	7	45
O_3	1 ㉚	12	4	7	7	30
O_4	10	15	6 ⑩	9	1 ㊵	50
Total	40	20	50	30	40	

optimum solution by the steppingstone method.   The data of Table 10.6 are reproduced in Table 10.12.

Check on Step 1

Since the number of occupied cells in this program equals $m + n - 1$, that is, $4 + 5 - 1 = 8$, this is indeed a basic feasible solution.

Step 2   Determine the Opportunity Costs of the Empty Cells

We repeat: in the steppingstone method a separate closed loop with proper plus and minus signs must be completed for *each* of the empty cells *before* the respective opportunity costs can be calculated.* Since our first program has a total of 12 empty cells, 12 different closed loops must be drawn.   The opportunity cost associated with each empty cell is calculated in Table 10.13.   An examination of these opportunity costs

---

* The reader should firmly grasp both the logic and the technique utilized in drawing these closed loops.   Note the difference between the closed loop for cell O_1D_2 and that for O_3D_2.

Table 10.13 Calculation of the Opportunity Costs for Table 10.6 and Indicated Changes

Empty cell	Closed loop	Net cost change	Opportunity cost	Action
O_1D_2	$+O_1D_2 - O_1D_3 + O_2D_3 - O_2D_2$	$+4 - 9 + 6 - 1 = 0$	0	Indifferent
O_1D_5	$+O_1D_5 - O_4D_5 + O_4D_3 - O_1D_3$	$+9 - 1 + 6 - 9 = +5$	-5	Do not include in next program
O_2D_1	$+O_2D_1 - O_1D_1 + O_1D_3 - O_2D_3$	$+8 - 12 + 9 - 6 = -1$	+1	Candidate for inclusion in next program
O_2D_4	$+O_2D_4 - O_1D_4 + O_1D_3 - O_2D_3$	$+6 - 5 + 9 - 6 = +4$	-4	Do not include in next program
O_2D_5	$+O_2D_5 - O_2D_3 + O_4D_3 - O_4D_5$	$+7 - 6 + 6 - 1 = +6$	-6	Do not include in next program
O_3D_2	$+O_3D_2 - O_3D_1 + O_1D_1 - O_1D_3 + O_2D_3 - O_2D_2$	$+12 - 1 + 12 - 9 + 6 - 1 = +19$	-19	Do not include in next program
O_3D_3	$+O_3D_3 - O_1D_3 + O_1D_1 - O_3D_1$	$+4 - 9 + 12 - 1 = +6$	-6	Do not include in next program
O_3D_4	$+O_3D_4 - O_1D_4 + O_1D_1 - O_3D_1$	$+7 - 5 + 12 - 1 = +13$	-13	Do not include in next program
O_3D_5	$+O_3D_5 - O_4D_5 + O_4D_3 - O_1D_3 + O_1D_1 - O_3D_1$	$+7 - 1 + 6 - 9 + 12 - 1 = +14$	-14	Do not include in next program
O_4D_1	$+O_4D_1 - O_1D_1 + O_1D_3 - O_4D_3$	$+10 - 12 + 9 - 6 = +1$	-1	Do not include in next program
O_4D_2	$+O_4D_2 - O_2D_2 + O_2D_3 - O_4D_3$	$+15 - 1 + 6 - 6 = +14$	-14	Do not include in next program
O_4D_4	$+O_4D_4 - O_4D_3 + O_1D_3 - O_1D_4$	$+9 - 6 + 9 - 5 = +7$	-7	Do not include in next program

shows that cell O_2D_1 is the only cell with a positive opportunity cost. Hence, cell O_2D_1 must be included in our next program.

A comment about the opportunity cost of zero (for empty cell O_1D_2) is in order. An opportunity cost of zero associated with a particular empty cell at any stage of the problem solution indicates that if this cell is included in the next program the total cost of the new program will be the same as that of the current program. Thus we list "Indifferent" in the "Action" column of Table 10.13.

Step 3   Revising a Given Program

Having ascertained that a given program is not an optimal program (one or more empty cells have positive opportunity cost), we next revise the

Table 10.14

First program

	D_1	D_2	D_3	D_4	D_5
O_1	12 ⑩	4	9 ⑮	5 ㉚	9
O_2	8 □	1 ⑳	6 ㉕	6	7
O_3	1 ㉚	12	4	7	7
O_4	10	15	6 ⑩	9	1 ㊵

Revised program

	D_1	D_2	D_3	D_4	D_5
O_1	12	4	9 ㉕	5 ㉚	9
O_2	8 ⑩	1 ⑳	6 ⑮	6	7
O_3	1 ㉚	12	4	7	7
O_4	10	15	6 ⑩	9	1 ㊵

given program to obtain a new basic feasible solution. The revised program must include that empty cell of the current program whose opportunity cost is highest. No choice is necessary here, since cell O_2D_1 is the only empty cell having a positive opportunity cost.

The revision of the first program is guided by the closed loop of the empty cell to be included (in this case cell O_2D_1) and is as shown in Table 10.14. Since 10 is the smallest number in a negative cell in the closed loop, it is added to the cells containing plus signs and subtracted from the cells containing minus signs.

The next question is: Does our revised program represent an optimal solution? To answer this question, we have to repeat step 2, as discussed previously. Should the result of step 2 indicate a nonoptimal solution, we would repeat step 3, namely, obtain another basic feasible solution. In other words, after the initial basic feasible solution has been obtained, the optimal solution is derived by the repeated application of steps 2 and 3. Determination of the opportunity costs of all the empty cells of the revised program (Table 10.14) will reveal that an optimal solution has indeed been derived. The reader is encouraged to verify this by the application of step 2 to Table 10.14.

10.8 MODIFIED-DISTRIBUTION METHOD FOR DERIVING AN OPTIMAL SOLUTION (MODI)

The main difference between the steppingstone method and the modified-distribution method (MODI) for solving transportation problems concerns the stage of the problem solution at which the closed loop(s) is drawn. In the steppingstone method, the closed loops for all the empty cells are drawn *before* their respective opportunity costs can be calculated. The empty cell to be included in the next program is then identified as that having the highest opportunity cost. In other words, the procedure for calculating the opportunity costs of the empty cells is dependent on the tracing of the closed loops.

In the modified-distribution method, however, the opportunity costs of all the empty cells are calculated and the highest opportunity cost is identified before any closed loop is drawn. As a matter of fact, in the modified-distribution method we need draw *only one* closed loop *after* the highest-opportunity-cost cell has been identified. Thus, the procedure for calculating the opportunity costs of the empty cells in MODI is independent of the tracing of the loops.

We shall illustrate the mechanics and rationale of the modified-distribution method by solving the simple transportation problem of Table 10.7, the data of which are reproduced in Table 10.15. The initial assignment made by following the northwest-corner rule is given in Table 10.16.

Determining the Opportunity Costs of the Empty Cells

Having solved this problem by the steppingstone method, we are aware of the fact that the transfer of 1 unit to cell O_2D_1 (a closed loop for the empty cell was established in Table 10.9 *before* this transfer was made) results in a net cost change of -1 dollar. This, of course, means that the opportunity cost of *not* including cell O_2D_1 in the first program is $+1$ dollar per unit of goods. Another method of reaching the same conclusion is via the determination of what may be called the *implied cost* of an empty cell.   The implied cost of an empty cell sets an upper limit (in

Table 10.15

Origin	Destination		Origin capacity per time period
	D_1	D_2	
O_1	2	2	1,000
O_2	1	2	600
Destination requirement per time period	900	700	

Table 10.16 *Initial Basic Feasible Solution*

	D_1	D_2
O_1	2 (900)	2 (100)
O_2	1	2 (600)

view of the existing program*) beyond which the inclusion of this cell in a new program is not an advantageous proposition.   Let us explain in connection with our transportation problem.

* The implied cost of a given empty cell can change from one program to another, since the implied cost is indicative of the *relative* advantage or disadvantage of not using a given cell in a particular program.

In this case, cell O_2D_1 is the only empty cell. One way to find its implied cost is by drawing a closed loop and determining the net cost consequence of shifting 1 unit of goods into O_2D_1. Ignoring, for the time being, the actual shipping cost per unit via route O_2D_1, we may calculate the net cost consequence of shifting 1 unit of goods into O_2D_1 as

$$O_2D_1 - O_1D_1 + O_1D_2 - O_2D_2 = O_2D_1 - 2 + 2 - 2 = \boxed{O_2D_1 - 2}$$ *negative*

Whatever the actual shipping cost per unit via cell O_2D_1, it is obvious that the above shift is desirable only if the net cost change $(O_2D_1 - 2)$ is negative. It will be negative so long as the actual cost of O_2D_1 is less than 2. The calculated upper limit for the actual cost of cell O_2D_1 (in the existing program), beyond which the inclusion of this cell is not an advantageous proposition, is therefore 2. In other words, if the actual shipping cost via cell O_2D_1 is greater than \$2 per unit, the shift is not desirable. On the other hand, if the actual shipping cost is less than \$2 per unit, the shift is desirable and cell O_2D_1 should be included in the next program.

The implied cost of the empty cell O_2D_1 therefore is \$2 per unit.

Also, as we noted earlier, the negative of the net cost change involved in shifting 1 unit of goods to an empty cell gives the opportunity cost associated with the empty cell. For cell O_2D_1,

Opportunity cost $= -$(net cost change) $= -(O_2D_1 - 2) = 2 - O_2D_1$

where O_2D_1 is the *actual* cost of shipment per unit via cell O_2D_1. But, as we have just calculated, the implied cost of not using cell O_2D_1 is \$2 per unit. Hence,

Opportunity cost $=$ implied cost $-$ actual cost

Substituting the actual shipping cost via cell O_2D_1 (\$1) and the calculated implied cost of cell O_2D_1 in the above expression, we find that the opportunity cost (of cell O_2D_1) is $2 - 1 = +1$ dollar. This is the same value of opportunity cost (for cell O_2D_1) that we found earlier by a direct observation of the net cost consequence associated with shifting 1 unit of goods into cell O_2D_1. This equivalence holds for *any* empty cell, and we state again the general relationship:

Opportunity cost $=$ implied cost $-$ actual cost

Although we have now succeeded in determining the opportunity cost of an empty cell by developing the concept of implied cost, it has been possible to do so only by first drawing a closed loop.   The next logical question is: Can we somehow determine the implied cost of an empty cell without first drawing the closed loop?   Should we find this to be possible, we would establish the main framework for the MODI method, for then we could subtract the actual cost of the empty cell from its calculated implied cost and thus determine its opportunity cost without first drawing the closed loop.*

In this and the following paragraphs we shall develop a method for determining the implied costs of empty cells without drawing their respective loops.   Let us refer back to the initial basic feasible solution of Table 10.16.   In this program we have three occupied cells.   In linear-programming terms, this means that three (x_{11}, x_{12}, x_{22}) of the four variables are basis variables.   It will be recalled from the simplex method that the opportunity cost (represented by the numbers in the net-evaluation row) of any variable comprising the basis is zero.   Similarly, it can be shown in the case of the transportation problem that the opportunity cost of each of the occupied cells (cells containing the basis variables) is zero.   In other words, if the basis variables are not going to be changed, then the hypothetical introduction and removal of 1 unit in any occupied cell will not result in any net cost change.   Now, if we assign a complete set of row numbers (to be placed at the extreme right-hand side of the table containing a given program) and a complete set of column numbers (to be placed at the bottom of the table) in such a way that the shipping cost per unit of *each* of the occupied cells equals the sum of its row and column numbers, we shall satisfy the condition that the opportunity cost of each occupied cell be zero.†   Further, since the sum of the row and column numbers of any occupied cell equals the cost of that cell (a basis variable), the sum of the row and column numbers corresponding to each empty cell (nonbasis routes) gives the implied cost of that empty cell.

---

* As the reader will recall, it is this feature that distinguishes the MODI method from the steppingstone method.

† The transportation problem, if fed into the first simplex tableau, consists of column vectors representing structural and other variables (see Figure 10.1).   In each column vector representing a structural variable, two of the entries are 1, and the rest are 0. It is this special property of the transportation problem which makes it quite easy to find a new basis by the MODI method.   The MODI method guides us to a new basis after all the empty cells of the transportation matrix have been "evaluated" simultaneously.   A set of row numbers u_i and a set of column numbers v_j are chosen so that the opportunity cost of *each* cell is given by $u_i + v_j - c_{ij}$, where c_{ij} is the actual shipping cost per unit of the cell falling in ith row and jth column.   *Thus, if we choose u_i and v_j such that for all the occupied cells (basis routes) $c_{ij} = u_i + v_j$, we satisfy the requirement that the opportunity cost of each occupied cell is zero.*   For the empty cells (nonbasis routes), opportunity cost is given by $u_i + v_j - c_{ij}$.

The implied cost of *any* empty cell, therefore, is given by

Implied cost = row number + column number = $u_i + v_j$

Thus, by the assignment of row and column numbers, we can calculate the implied cost of each empty cell without drawing a closed loop. We must now tackle the problem of assigning these row and column numbers.

For each occupied cell, we have to choose u_i (row number) and v_j (column number) such that c_{ij} (the actual shipping cost per unit in the occupied cell) equals the sum of u_i and v_j. For the occupied cell falling in row 1 and column 1, for example, u_1 and v_1 are chosen such that $c_{11} = u_1 + v_1$. Similarly, for cell O_1D_2 we must chose u_1 and v_2 such that $c_{12} = u_1 + v_2$. This process must be carried out for all the occupied cells. But it should be realized that although a basic feasible solution for a transportation problem consists of $m + n - 1$ variables (in other words, there are $m + n - 1$ occupied cells), we must assign $m + n$ values to obtain a complete set of row and column numbers. Hence, to determine all the row and column numbers, one *arbitrary* number, serving as either a row or a column number, must be chosen. Once one row number or column number has been chosen arbitrarily, the rest of the row and column numbers can be determined by the relationship $c_{ij} = u_i + v_j$. This relationship, as stated earlier, *must* hold for all the *occupied* cells. In so far as any arbitrary number can be chosen to represent one of the u_i's or v_j's, we shall follow the practice of making u_1 take the value zero. This completes the description of the procedure for determining the row and column numbers. The actual numbers for our example are given in Table 10.17. If we arbitrarily choose a value of zero for u_1, our next question is: What value must be given to v_1 so that $c_{11} = u_1 + v_1$ or $2 = 0 + v_1$? Obviously, v_1 must take a value of 2. Next, we ask: What value must be given to v_2 so that $c_{12} = u_1 + v_2$ or $2 = 0 + v_2$? The value of v_2 must be 2. Again, what value must be given to u_2 so that $c_{22} = u_2 + v_2$ or $2 = u_2 + 2$? Obviously, $u_2 = 0$. By first assigning an arbitrary value to u_1 and then posing a series of questions, we have determined all the row and column numbers.

Let us now calculate the opportunity cost for the empty cell O_2D_1. The opportunity cost of an empty cell, as stated earlier, is given by implied cost − actual cost, that is, by $(u_i + v_j) - c_{ij}$. For cell O_2D_1, therefore, the opportunity cost is $u_2 + v_1 - c_{21} = 0 + 2 - 1 = +1$ dollar. The answer, of course, is the same as that obtained by the long method. In so far as the opportunity cost of the empty cell O_2D_1 is positive, this is not an optimum program and hence must be revised.

Before we revise the above program, let us summarize the role of the row and column numbers.   In so far as the row and column numbers are assigned in such a manner that the *actual* cost of every occupied cell equals the sum of its row and column numbers, the sum of the row and column numbers of each empty cell gives the implied cost of that empty

Table 10.17

Origin	Destination		Row number
	D_1	D_2	
O_1	[2] 〈900〉	[2] 〈100〉	0
O_2	[1]	[2] 〈600〉	0
Column number	2	2	

Table 10.18

Implied cost		Actual cost	Action
$u_i + v_j$	>	c_{ij}	A better program can be designed by including this cell in the solution
$u_i + v_j$	=	c_{ij}	Indifferent; however, an alternative program with the same total cost and including this cell can be designed
$u_i + v_j$	<	c_{ij}	Do not include this cell in the program

cell (unused route).   If the implied cost of the empty cell is less than its actual cost, this route should be left out of our shipping program.   If, on the other hand, the implied cost $(u_i + v_j)$ of an empty cell is more than its actual cost (c_{ij}), then this route would be a candidate for inclusion in our next program.   In summary, to evaluate and improve a given program in which the objective is to *minimize* a given function, the rules given in Table 10.18 apply.   For a transportation problem in which the objective is to *maximize* a given function, the signs of the inequalities given in the table must be reversed to establish the guidelines for action. Let us now return to our problem.

Revising a Given Program

The last step in the MODI method is exactly the same as the corresponding step in the steppingstone method.   Having identified the empty cell to be included in the next program (the cell with the highest opportunity cost), we draw a closed loop for this cell.   The new basic feasible solution is then derived by shifting into the empty cell the maximum possible

Table 10.19

First program   *Revised program*

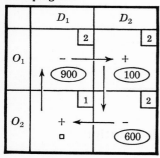

number of units without violating the rim requirements.   The revised program is given in Table 10.19.   To determine if the revised program is an optimal program, we must determine the opportunity cost of the empty cell O_2D_2.   This is illustrated in Table 10.20.   From the table, we see that

$$\text{\textit{Implied} cost of cell } O_2D_2 = u_2 + v_2 = -1 + 2 = +1$$

$$\text{Actual cost of cell } O_2D_2 = +2$$

Hence

Opportunity cost of empty cell O_2D_2 = implied cost − actual cost

$$= +1 - 2 = -1$$

In so far as the opportunity cost of the only empty cell is nonpositive, no improvement in the present program is possible.   This program, then, is the optimal program.

We shall now apply the modified-distribution method to the problem of Table 10.2.

Table 10.20

Origin	Destination		Row number
	D_1	D_2	
O_1	[2] (300)	[2] (700)	0
O_2	[1] (600)	[2]	-1
Column number	2	2	

Step 1   Obtaining an Initial Basic Feasible Solution

As discussed earlier, an initial basic feasible solution for a given transportation problem may be obtained by following the northwest-corner rule, by the application of Vogel's approximation method, or simply by inspection.   Table 10.21 reproduces the basic feasible solution of Table 10.6 (initial assignment by inspection), which we shall take as a starting point for obtaining the optimal solution by MODI.   In so far as the number of occupied cells in this program equals $m + n - 1$, that is, $4 + 5 - 1 = 8$, this is indeed a basic feasible solution.

Step 2   Determining the Opportunity Costs of the Empty Cells

To determine the opportunity costs of the empty cells by the MODI method, we must first determine the implied costs of the empty cells by assigning a complete set of row and column numbers.   This is shown in Table 10.22.   The uncircled numbers in the matrix represent the implied

Table 10.21 *First Program*

Origin	Destination					Total
	D_1	D_2	D_3	D_4	D_5	
O_1	12 (10)	4	9 (15)	5 (30)	9	55 0
O_2	8	1 (20)	6 (25)	6	7	45 −3
O_3	1 (30)	12	4	7	7	30 −11
O_4	10	15	6 (10)	9	1 (40)	50 −3
Total	40 *12*	20 *4*	50 *9*	30 *5*	40 *4*	

Table 10.22

	D_1	D_2	D_3	D_4	D_5	Row number
O_1	12 (10)	4 4	9 (15)	5 (30)	9 4	0
O_2	8 9 *+1*	1 (20)	6 (25)	6 2 *−4*	7 1 *−6*	−3
O_3	1 (30)	12 −7	4 −2 *−6*	7 −6 *−13*	7 −7 *−14*	−11
O_4	10 9 *−1*	15 1 *−14*	6 (10)	9 2 *−7*	1 (40)	−3
Column number	12	4	9	5	4	

costs of the empty cells.   A comparison of the implied and actual costs of each empty cell shows that only cell O_2D_1 has a positive opportunity cost of $+1$ dollar.   For cell O_2D_1, opportunity cost $=$ implied cost $-$ actual cost $= 9 - 8 = +1$.   A similar calculation for cell O_1D_2 shows that its opportunity cost is zero.   The opportunity costs for the rest of the empty cells are negative.

Having identified the presence of positive opportunity cost, we know that this program is not an optimum program.   Hence it must be revised to include that empty cell which has the highest opportunity cost (in this case cell O_2D_1).

Step 3   Revising the Given Program

The revision of the given program is guided by a closed loop drawn for the empty cell which is to be included in the next program.   The loop for

Table 10.23

	D_1	D_2	D_3	D_4	D_5	Row number
O_1	12 11	4 4	9 (25)	5 (30)	9 4	0
O_2	8 (10)	1 (20)	6 (15)	6 2	7 1	-3
O_3	1 (30)	12 -6	4 -1	7 -5	7 -6	-10
O_4	10 8	15 1	6 (10)	9 2	1 (40)	-3
Column number	11	4	9	5	4	

cell O_2D_1 is drawn and the program is revised in exactly the same manner as shown in Table 10.14.   The revised program is then tested for optimality (by the application of step 2), as shown in Table 10.23.   A comparison of the uncircled numbers (representing the implied costs) in the empty cells and the respective actual costs shows that no empty cell has a positive opportunity cost.*   Hence, this is an optimal solution.

10.9 PROCEDURE SUMMARY FOR THE MODIFIED-DISTRIBUTION METHOD (MINIMIZATION CASE)

Step 1 Obtain a Basic Feasible Solution

An initial basic feasible solution for a given transportation problem may be obtained by following the northwest-corner rule, by the application of Vogel's approximation method, or by simple inspection.

Test for step 1.   A basic feasible solution must include shipments covering $m + n - 1$ cells.   That is, the number of occupied cells (basis variables) is 1 less than the number of rows and columns in the transportation matrix.

If the number of occupied cells in the initial solution is more than $m + n - 1$, there is a computational error which can easily be corrected by rechecking the data.   If the number of occupied cells is less than $m + n - 1$, this is a degenerate solution.   To resolve degeneracy, add one or more epsilons to some "suitable" empty cells so that the number of occupied cells becomes equal to $m + n - 1$.†

Step 2 Determine the Opportunity Costs of the Empty Cells (*Opportunity Cost = Implied Cost − Actual Cost*)

a. *Determine a complete set of row and column numbers (values).*   When, in a given program, the number of occupied cells equals $m + n - 1$, proceed to assign row and column numbers (values) in such a manner that, for *each occupied cell,* the relationship $c_{ij} = u_i + v_j$ holds.   To start, a value of zero can be assigned to any row having an occupied cell.   The rest of the row and column numbers can then be determined by making sure that, for each occupied cell, $c_{ij} = u_i + v_j$.   In other words, for each occupied cell, the actual shipping cost per unit should equal the sum of its row and column values.

b. *Calculate the implied costs of the empty cells.*   Once all the row and column values have been assigned, the implied cost of a given empty

---

* The fact that empty cell O_1D_2 has an opportunity cost of zero means that an alternative program which will include cell O_1D_2 and have the same total shipping cost as this program can be designed.

† Degeneracy in transportation problems is illustrated in Section 10.13.

cell can be calculated as follows:

Implied cost = row value + column value

c. *Determine the opportunity costs of the empty cells.* The opportunity cost of an empty cell is determined by subtracting the actual cost of the empty cell from its implied cost. In other words, opportunity cost, for each cell, is given by

$$\text{Opportunity cost} = \overset{\text{implied cost}}{u_i + v_j} - \overset{\text{actual cost}}{c_{ij}}$$

If the opportunity costs of all the empty cells are nonpositive, an optimal solution has been obtained. If, on the other hand, even a single cell has a positive opportunity cost, a better program can be designed. Thus, step 2 serves as a test for optimality.

Step 3   Design an Improved Program

Design a new program such that the empty cell having the largest opportunity cost (in the program to be revised) is included in the solution. This is accomplished in the following manner:

a. Draw a loop of horizontal and vertical arrows in such a manner that it starts from the empty cell to be filled, passes to the nearest occupied cell in the same row or column as the empty cell, and then, making a series of alternate horizontal and vertical turns through *occupied* cells, returns to the original empty cell.
b. Place a plus sign (+) in the empty cell to be filled. Then, alternately, place minus signs (−) and plus signs (+) at the beginnings and ends of the connecting links of the loop.
c. Examine those occupied cells in which the minus signs have been placed. Of these, the cell having the least number of units is vacated by transferring these units to the empty cell. This is accomplished by adding the same amount to all cells having plus signs and subtracting it from all cells having minus signs. The improved program should have the same number of occupied cells as the preceding program. If the number of occupied cells in the improved program is less than that of the preceding program, the problem becomes degenerate. In such a case, add epsilon(s) to some *recently vacated* cell(s) such that the number of occupied cells again equals $m + n - 1$.*

* See Section 10.13.

Step 4

Repeat steps 2 and 3 until a program is achieved in which each empty cell has an opportunity-cost value which is either zero or negative. This program will be the optimal program.

10.10 MODIFIED-DISTRIBUTION METHOD (MAXIMIZATION CASE)

Except for one transformation, a transportation problem in which the objective is to maximize a given function can be solved by the MODI algorithm as presented above. The transformation is made by subtracting all the c_{ij}'s from the highest c_{ij} (profit) of the given transportation matrix. The transformed c_{ij}'s give us the *relative costs*, and the problem then becomes a minimization problem. Once an optimal solution to this transformed minimization problem has been found, the value of the objective function can be calculated by inserting the original values of the c_{ij}'s for those routes which form the basis (occupied cells) in the optimal solution.

10.11 BALANCING THE GIVEN TRANSPORTATION PROBLEM

To solve a given transportation problem by the step-by-step procedure given in Section 10.9, we must establish equality between the total capacities of the origins and the total requirements of the destinations. Three cases can arise.

Case 1 $\Sigma b_i = \Sigma d_j$

In this case, the total capacity of the origins equals the total requirement of the destinations. The problem can be arranged in the form of a matrix, along with the relevant cost data, and the transportation algorithm may be applied directly to obtain a solution.

Case 2 $\Sigma b_i > \Sigma d_j$

In this case, the total capacity of the origins exceeds the total requirement of the destinations. A "dummy" destination can be added to the matrix to absorb the excess capacity.   The cost of shipping from each origin to this dummy destination is assumed to be zero.   The adding of a dummy destination establishes equality between the total origin capacities and total destination requirements.   The problem is then amenable to solution by the transportation algorithm.

Illustrative Example

Table 10.24 gives both the unbalanced and balanced forms of a transportation problem in which the total given capacity of the origins exceeds the total given requirement of the destinations ($\Sigma b_i > \Sigma d_j$).   In practice.

Table 10.24

Unbalanced form

	D_1	D_2	D_3	Origin capacity
O_1	5	3	2	200
O_2	6	4	1	400
Destination requirement	200	200	150	

Balanced form

	D_1	D_2	D_3	Dummy	Origin capacity
O_1	5	3	2	0	200
O_2	6	4	1	0	400
Destination requirement	200	200	150	50	

the optimal solution identifies the particular origin at which the excess capacity should be left idle.

Case 3 $\Sigma b_i < \Sigma d_j$

In this case, the total capacity of the origins is less than the total requirement of the destinations.   A dummy origin can be added to the transportation matrix to meet the excess demand.   The cost of shipping from the dummy origin to each destination is assumed to be zero.   The adding of a dummy origin in this case establishes the equality between the total capacity of the origins and the total requirement of the destinations.*

---

* The reader will observe that the role of the dummy column or dummy row containing dummy variables in a transportation problem is parallel to the role of the slack variables in the general linear-programming problems illustrated previously.

Illustrative Example

Table 10.25 gives both the unbalanced and balanced forms of a transportation problem in which the total given capacity of the origins is less than the total requirement of the destinations ($\Sigma b_i < \Sigma d_j$). In practice, the optimal solution identifies the particular destination whose requirement cannot be fully satisfied.

In the initial assignment for a transportation problem which has been balanced by the addition of a dummy origin or a dummy destination, only the last necessary allocations should be made to the dummy cells.

Table 10.25

Unbalanced form

	D_1	D_2	D_3	Origin capacity
O_1	[5]	[3]	[2]	200
O_2	[6]	[4]	[1]	400
Destination requirement	300	200	150	

Balanced form

	D_1	D_2	D_3	Origin capacity
O_1	[5]	[3]	[2]	200
O_2	[6]	[4]	[1]	400
Dummy	[0]	[0]	[0]	50
Destination requirement	300	200	150	

This procedure, in general, will result in fewer iterations before an optimal solution is derived.

10.12   SIMPLEX TRANSLATION OF THE TRANSPORTATION METHOD

Some important observations can be made regarding parallels in the general transportation model and the simplex method. First, the role of the dummy variables in the transportation problem is similar to the role of the slack variables in the general linear-programming problem. Second, the occupied cells and empty cells of the transportation program correspond, respectively, to the basis variables and nonbasis variables of the simplex tableau.

Third, the revision of a given transportation program is parallel to the process of obtaining a new basis in the simplex method. Let us explain this point further.   A given transportation program, it will be recalled, is improved by filling or including one empty cell (that having the highest opportunity cost) at a time.   In this process, all the units from *at least one* cell are removed.   Thus, a new cell is filled and becomes an occupied cell, and at least one of the previously occupied cells joins the category of empty cells.   The total number of occupied cells, therefore, can either remain constant (only one previously occupied cell becomes an empty cell) or decrease (more than one of the previously occupied cells become empty cells) from one program to the next.   If the number of occupied cells remains the same from one program to the next, the process is similar to a simplex iteration in which one new (nonbasis) variable is introduced into the solution to remove one of the basis variables currently in the solution.   Of course, in this case we obtain a new basic feasible solution.   If, on the other hand, the process of filling one empty cell results in the simultaneous vacating of two or more of the currently occupied cells, the transportation problem becomes degenerate.   This latter situation, as the reader will observe, is parallel to the simplex iteration in which the introduction of one new (nonbasis) variable removes, simultaneously, two or more of the current basis variables— here, too, the problem becomes degenerate.

10.13   DEGENERACY IN TRANSPORTATION PROBLEMS

It was established earlier that a basic feasible solution for a transportation problem consists of $m + n - 1$ basis variables.   This means that the number of occupied cells in a given transportation program is 1 less than the number of rows and columns in the transportation matrix.   Whenever the number of occupied cells is less than $m + n - 1$, the transportation problem is said to be degenerate.

Degeneracy in transportation problems can develop in two ways. First, the problem may become degenerate when the initial program is designed via one of the initial-assignment methods discussed earlier.   To resolve degeneracy in this case, we can allocate an extremely small amount of goods (close to zero) to one or more of the empty cells,* so that the number of occupied cells becomes $m + n - 1$.   The cell containing this extremely small allocation is, of course, considered to be an occupied cell.

---

* This extremely small amount, represented by epsilon, ϵ, may be allocated to any empty cell subject to the condition that this will make possible the determination of a unique set of row and column numbers.

In linear-programming literature, this extremely small amount is usually denoted by the Greek letter ϵ (epsilon). The amount ϵ is assumed to be so small that its addition to or subtraction from a given number does not change that number. For example, $50 + \epsilon = 50$, and $200 - \epsilon = 200$. Of course, if ϵ is subtracted from itself, the result is assumed to be zero; that is, $\epsilon - \epsilon = 0$.

The development of degeneracy during the initial assignment and its resolution will be illustrated with the transportation problem of Table 10.26.

Table 10.26 *Data for the Transportation Problem*

Origin	Destination			Origin capacity
	D_1	D_2	D_3	
O_1	2	1	2	20
O_2	3	4	1	40
Destination requirement	20	15	25	

Second, the transportation problem may become degenerate during the solution stages. This happens when the inclusion of the most favorable empty cell (the cell having the highest opportunity cost) results in the simultaneous vacating of two or more of the currently occupied cells. To resolve degeneracy in this case, we allocate ϵ to one or more of the recently vacated cells, so that the number of occupied cells in the new program is $m + n - 1$. This type of degeneracy and its resolution will be illustrated with the transportation problem of Table 10.30.

Case 1 Degeneracy during the Initial Assignment

Following the northwest-corner rule, we obtain the initial assignment given in Table 10.27. Note that the number of occupied cells in this program is 3, which does not equal $m + n - 1$.   Hence, the problem is degenerate at the very beginning, and no attempt to assign row and column numbers to Table 10.27 will succeed.   However, we can resolve this degeneracy by the addition of epsilon to any of the empty cells.   In so far as this is a minimization problem, we allocate epsilon to the lowest-cost cell O_1D_2 (see Table 10.28).   With this modification, the number of occupied cells equals 4, that is, $m + n - 1$.   Hence, it is now possible to assign a unique set of row and column numbers in order to apply the MODI method (see Table 10.28).   It is clear that the implied cost of cell O_2D_1 is 5 ($u_2 + v_1 = 2 + 3 = 5$), whereas its actual cost is 3.   Hence,

Table 10.27   *Initial Assignment by Northwest-corner Rule (a Degenerate Solution)*

	D_1	D_2	D_3	Total
O_1	2 ⑳	1	2	20
O_2	3	4 ⑮	1 ㉕	40
Total	20	15	25	

the opportunity cost of cell O_2D_1 is $5 - 3 = +2$.   Thus, the program in Table 10.28 is not optimal.

The revision of the program is shown in Table 10.29.   The closed loop of the table shows that at most 15 units can be shifted to cell O_2D_1. This necessitates the subtraction of 15 units from O_1D_1 and from O_2D_2 and the addition of 15 units to O_1D_2.   In so far as $15 + \epsilon = 15$, cell O_1D_2 has been assigned a total of 15 units by this shifting process.   Also, note that the revised program is not a degenerate solution.   If we test the revised program for optimality, we find that it represents an optimal solution to the given problem.   The reader can immediately verify this by assigning a set of row and column numbers to the matrix representing the revised program and then calculating the opportunity costs of its empty cells.

When degeneracy developed in the initial assignment for the above transportation problem (Table 10.27), we noted that the addition of ϵ to any of the empty cells enabled us to determine a unique set of row and column numbers. This is usually true whenever the initial assignment is made by following the northwest-corner rule. However, when the initial assignment is made by another method, such as by inspection, one cannot add epsilon to just any empty cell. Instead, ϵ must be added to one of

Table 10.28*

	D_1	D_2	D_3	Row number
O_1	2 ㉒	1 ⓔ	2	0
O_2	3	4 ⑮	1 ㉕	3
Column number	2	1	-2	

* Initial basic feasible solution (after the addition of ϵ to cell O_1D_2) and its row and column numbers.

Table 10.29

First program

	D_1	D_2	D_3	Total
O_1	2 ㉒	1 ⓔ	2	20
O_2	3 ▫	4 ⑮	1 ㉕	40
Total	20	15	25	

Revised program

	D_1	D_2	D_3	Total
O_1	2 ⑤	1 ⑮	2	20
O_2	3 ⑮	4	1 ㉕	40
Total	20	15	25	

those empty cells which will make possible the determination of a unique set of row and column numbers. Let us illustrate this point by considering one particular degenerate solution of the transportation problem of Table 10.30.

Clearly, the initial assignment, by inspection, gives a degenerate solution, since the number of occupied cells is less than $m + n - 1$. To resolve this degenerate solution which has developed at the very start, we must add ϵ to one of the empty cells. But we must choose this empty cell with careful judgment, for if we make any one of the empty cells O_3D_2, O_4D_1, and O_5D_1 an occupied cell by the addition of ϵ we shall not be able to assign a unique set of row and column numbers. On the other hand, the addition of ϵ to any one of the cells O_1D_1, O_1D_2, O_2D_3, O_3D_3, O_4D_3, and O_5D_3 will enable us to resolve the degeneracy and allow us to determine a unique set of row and column numbers. The task of verifying these statements is left to the reader.

Table 10.30

Data for the transportation problem

An initial assignment by inspection

	D_1	D_2	D_3	Total
O_1	[2]	[4]	[1]	40
O_2	[6]	[3]	[2]	50
O_3	[4]	[5]	[6]	20
O_4	[3]	[2]	[1]	30
O_5	[5]	[2]	[5]	10
Total	50	60	40	

	D_1	D_2	D_3	Total
O_1	[2]	[4]	[1] (40)	40
O_2	[6] (30)	[3] (20)	[2]	50
O_3	[4] (20)	[5]	[6]	20
O_4	[3]	[2] (30)	[1]	30
O_5	[5]	[2] (10)	[5]	10
Total	50	60	40	

Once a unique set of row and column numbers has been determined, the various steps of the transportation algorithm can be applied in a routine manner to obtain an optimal solution.

Case 2   Degeneracy during the Solution Stages

Given in Table 10.31a and b are a transportation problem and an initial assignment derived by following the northwest-corner rule. The solution represented by Table 10.31b is a basic feasible solution. Should we assign a set of row and column numbers to the empty cells of this program, we would discover that there are several empty cells (including cell O_1D_3) having positive opportunity costs. Let us decide, arbitrarily, to include cell O_1D_3 in a new program. This necessitates shifting 10 units to cell O_1D_3 as guided by the closed loop shown in Table 10.31b. The resulting program, given in Table 10.31c, is a degenerate solution.

Table 10.31a   Data for the Transportation Problem

	D_1	D_2	D_3	D_4	D_5	Total
O_1	4	3	1	2	6	40
O_2	5	2	3	4	5	30
O_3	3	5	6	3	2	20
O_4	2	4	4	5	3	10
Total	30	30	15	20	5	

Table 10.31b   Initial Assignment by Northwest-corner Rule

	D_1	D_2	D_3	D_4	D_5	Total
O_1	4 (30)	3 (10) + □	1	2	6	40
O_2	5 +	2 (20)	3 (10)	4	5	30
O_3	3	5 (5)	6 (15)	3	2	20
O_4	2	4	4	5 (5)	3 (5)	10
Total	30	30	15	20	5	

Table 10.31c   Program 2

	D_1	D_2	D_3	D_4	D_5	Total
O_1	4 (30)	3	1 (10)	2	6	40
O_2	5	2 (30)	3	4	5	30
O_3	3	5	6 (5)	3 (15)	2	20
O_4	2	4	4	5 (5)	3 (5)	10
Total	30	30	15	20	5	

Since the degeneracy has developed during the solution stages, we should resolve it by adding ϵ to one of the recently vacated cells, i.e., cell O_1D_2 or cell O_2D_3. The reader should verify that, in this case, a set of row and column values can be determined *only* if epsilon is added to one of the empty cells O_1D_2, O_3D_2, O_4D_2, O_2D_1, O_2D_3, O_2D_4, or O_2D_5.

In so far as this is a minimization problem, we should add epsilon to that recently vacated cell which has the lowest shipping cost per unit. There being a tie in this case between cell O_1D_2 and cell O_2D_3, we arbitrarily decide to add ϵ to cell O_1D_2. This enables us to assign a unique set of row and column values to program 2, calculate the opportunity costs of the empty cells, and identify the most favorable empty cell to be included in the next program (see Table 10.31e).

The solution represented by program 3 is, again, degenerate. We therefore add another epsilon to cell O_4D_4 because, of the two cells recently vacated (O_3D_3 and O_4D_4), cell O_4D_4 has the lower shipping cost per unit.* This means that we now have two cells (O_1D_2 and O_4D_4 in Table 10.31f) having an assignment of epsilon. This modification enables

Table 10.31d   *Revised Program 2*

	D_1	D_2	D_3	D_4	D_5	Row number
O_1	[4] (30)	[3] (ε)	[1] (10)	[2]	[6]	0
O_2	[5]	[2] (30)	[3]	[4]	[5]	-1
O_3	[3]	[5]	[6] (5)	[3] (15)	[2]	5
O_4	[2]	[4]	[4]	[5] (5)	[3] (5)	7
Column number	4	3	1	-2	-4	

Table 10.31e   *Program 3*

	D_1	D_2	D_3	D_4	D_5	Row number
O_1	[4] (25)	[3] (ε)	[1] (15)	[2]	[6]	
O_2	[5]	[2] (30)	[3]	[4]	[5]	
O_3	[3]	[5]	[6]	[3] (20)	[2]	
O_4	[2] (5)	[4]	[4]	[5]	[3] (5)	
Column number						

us to continue the application of the transportation algorithm. The assignment of row and column numbers to program 3 is shown in Table 10.31f. The calculation of the opportunity costs of the empty cells of program 3 will show that cell O_1D_4 is the most favorable cell. Hence, we draw a closed loop for cell O_1D_4 and include this cell in our next program (see Table 10.31g). Note that program 4 is essentially the same as program 3, except that one of the epsilons has been shifted from cell O_4D_4 (in program 3) to cell O_1D_4 (in program 4). This illustrates the procedure by which we handle the shifting process when the most negative amount shown in the negative terminals of the closed loop is epsilon (see the closed loop $O_1D_4 - O_4D_4 + O_4D_1 - O_1D_1$ in the revised program 3).

Calculation of the opportunity costs of the empty cells of program 4

* The reader should verify this and other details which have not been explicitly stated in the solution of this problem.

will reveal that cell O_3D_5 is the most favorable cell. Hence, we draw a closed loop for cell O_3D_5 and include this cell in our next program (see Table 10.31h). Notice that, as a result of the inclusion of cell O_3D_5 in program 5, cell O_1D_4 has become a normally occupied cell. That is, instead of having an assignment of ϵ, as was the case in program 4, cell

Table 10.31f *Revised Program 3*

	D_1	D_2	D_3	D_4	D_5	Row number
O_1	[4] (25)	[3] (ε)	[1] (15)	[2] □	[6]	0
O_2	[5]	[2] (30)	[3]	[4]	[5]	−1
O_3	[3]	[5]	[6]	[3] (20)	[2]	−4
O_4	[2] (5)	[4]	[4]	[5] (ε)	[3] (5)	−2
Column number	4	3	1	7	5	

Table 10.31g *Program 4*

	D_1	D_2	D_3	D_4	D_5	Row number
O_1	[4] (25)	[3] (ε)	[1] (15)	[2] (ε)	[6]	
O_2	[5]	[2] (30)	[3]	[4]	[5]	
O_3	[3]	[5]	[6]	[3] (20)	[2]	
O_4	[2] (5)	[4]	[4]	[5]	[3] (5)	
Column number						

Table 10.31h *Revised Program 4*

	D_1	D_2	D_3	D_4	D_5	Row number
O_1	[4] (25)	[3] (ε)	[1] (15)	[2] (ε)	[6]	0
O_2	[5]	[2] (30)	[3]	[4]	[5]	−1
O_3	[3]	[5]	[6]	[3] (20)	[2] □	1
O_4	[2] (5)	[4]	[4]	[5] (5)	[3]	−2
Column number	4	3	1	2	5	

Table 10.31i *Program 5*

	D_1	D_2	D_3	D_4	D_5	Row number
O_1	[4] (20)	[3] (ε)	[1] (15)	[2] (5)	[6]	
O_2	[5]	[2] (30)	[3]	[4]	[5]	
O_3	[3]	[5]	[6]	[3] (15)	[2] (5)	
O_4	[2] (10)	[4]	[4]	[5]	[3]	
Column number						

O_1D_4 now has an assignment of 5 units. The epsilons disappear during the solution stages of any transportation problem whose optimal program is a basic feasible solution (see Tables 10.27 through 10.29). However, if the optimal program of a given transportation problem is not a basic feasible solution, one or more epsilons will remain in the optimal program (see Tables 10.31e through 10.31i). In the latter case, we simply disregard the routes in which the epsilons appear.

A test of optimality applied to program 5 will show that it is not an optimal program. We therefore must design a new and better program (see Table 10.31*j*). The reader should verify the details of the derivation of program 6.

If we test program 6 for optimality, we shall find that it represents an optimal solution. The assignment of ϵ to cell O_1D_2 can now be disregarded, which means that no goods will be shipped from origin O_1 to destination D_2. It is to be observed that program 6 is an example of an optimal solution which is not a basic feasible solution.

Table 10.31*j*   Revised Program 5

	D_1	D_2	D_3	D_4	D_5	Row number
O_1	[4] (20)	[3] (ε)	[1] (15)	[2] (5)	[6]	0
O_2	[5]	[2] (30)	[3]	[4]	[5]	−1
O_3	[3] □	[5]	[6]	[3] (15)	[2] (5)	1
O_4	[2] (10)	[4]	[4]	[5]	[3]	−2
Column number	4	3	1	2	1	

Table 10.31*k*   Program 6

	D_1	D_2	D_3	D_4	D_5	Row number
O_1	[4] (5)	[3] (ε)	[1] (15)	[2] (20)	[6]	
O_2	[5]	[2] (30)	[3]	[4]	[5]	
O_3	[3] (15)	[5]	[6]	[3]	[2] (5)	
O_4	[2] (10)	[4]	[4]	[5]	[3]	
Column number						

10.14   ALTERNATIVE OPTIMAL SOLUTIONS TO TRANSPORTATION PROBLEMS

An optimal solution to a given transportation problem is not always a unique solution. The existence of more than one optimal solution for a transportation problem can be determined by examining the opportunity costs of the empty cells in the optimal program designed by following the transportation algorithm. If there is any empty cell having an opportunity cost of zero in the optimal program, another optimal program with the same total shipping cost as the first can always be designed. The second optimal program is obtained by revising the first program so as to include the zero-opportunity-cost cell. Let us illustrate this by considering the transportation problem of Table 10.2, for which the optimal solution in Table 10.23 was obtained. Along with a unique set of row and column numbers, the optimal solution of Table 10.23 is reproduced in Table 10.32.

The uncircled numbers in the matrix represent the implied costs of the empty cells. A quick visual check reveals that the opportunity cost of the empty cell O_1D_2 is zero.* Hence, the total shipping cost of a new program including cell O_1D_2 will be the same as the total shipping cost of

Table 10.32 *Matrix A (First Optimal Program)*

	D_1	D_2	D_3	D_4	D_5	Row number
O_1	11 [12]	4 [4]	(25) [9]	(30) [5]	4 [9]	0
O_2	(10) [8]	(20) [1]	(15) [6]	2 [6]	1 [7]	−3
O_3	(30) [1]	−6 [12]	−1 [4]	−5 [7]	−6 [7]	−10
O_4	8 [10]	1 [15]	(10) [6]	2 [9]	(40) [1]	−3
Column number	11	4	9	5	4	

Table 10.33

Matrix A (first optimum program)

	D_1	D_2	D_3	D_4	D_5	Total
O_1	[12]	[4]	(25) [9]	(30) [5]	[9]	55
O_2	(10) [8]	(20) [1]	(15) [6]	[6]	[7]	45
O_3	(30) [1]	[12]	[4]	[7]	[7]	30
O_4	[10]	[15]	(10) [6]	(40) [9]	[1]	50
Total	40	20	50	30	40	

Matrix B (second optimum program)

	D_1	D_2	D_3	D_4	D_5	Total
O_1	[12]	(20) [4]	(5) [9]	(30) [5]	[9]	55
O_2	(10) [8]	[1]	(35) [6]	[6]	[7]	45
O_3	(30) [1]	[12]	[4]	[7]	[7]	30
O_4	[10]	[15]	(10) [6]	(40) [9]	[1]	50
Total	40	20	50	30	40	

the present (first optimal) program. The revision of the first optimal program is illustrated in Table 10.33. The total shipping cost in each of the two programs is $695, as the reader can verify.

* Opportunity cost = implied cost − actual cost. For cell O_1D_2, the implied cost equals 4, and the actual cost equals 4. Thus, the opportunity cost of cell O_1D_2 is zero.

Once the existence of two alternative optimal programs is established, an infinite number of other alternative optimal programs can be derived. The following relationship governs the derivation of these alternative programs:

Derived program $= dA + (1 - d)B$

where A = matrix representing first optimal program
B = matrix representing second optimal program
d = any positive fraction less than 1

Let us illustrate the application of the above formula by considering the two optimal programs given in Table 10.33.   Assume that $d = \frac{2}{5}$.   Then

Derived program $= \frac{2}{5}A + \frac{3}{5}B$

Thus, we multiply each assignment in matrix A by $\frac{2}{5}$, multiply each assignment in matrix B by $\frac{3}{5}$, and add the corresponding elements (circled numbers representing allocations) of the two matrices.   The

Table 10.34 *Derived Program (Third Alternative Solution)*

	D_1	D_2	D_3	D_4	D_5	Total
O_1	[12]	[4] $0+12=\textcircled{12}$	[9] $10+3=\textcircled{13}$	[5] $12+18=\textcircled{30}$	[9]	55
O_2	[8] $4+6=\textcircled{10}$	[1] $8+0=\textcircled{8}$	[6] $6+21=\textcircled{27}$	[6]	[7]	45
O_3	[1] $12+18=\textcircled{30}$	[12]	[4]	[7]	[7]	30
O_4	[10]	[15]	[6] $4+6=\textcircled{10}$	[9]	[1] $16+24=\textcircled{40}$	50
	40	20	50	30	40	

result is the derived program given in Table 10.34. The total shipping cost of the derived program, as can easily be verified from Table 10.34, is the same ($695) as that of the first two optimal programs. The derived program fully satisfies the rim requirements. However, the number of occupied cells in the derived program is 9, as compared with 8 in the first two optimal programs. This means that our derived program is a feasible solution, but not a basic feasible solution.

In so far as we can let d be any positive fraction, it is obvious that an infinite number of derived solutions can be obtained so long as two alternative optimal solutions can be identified.

In terms of practical significance, the possibility of designing alternative solutions gives valuable flexibility to the decision maker. It should also be realized that an examination of the opportunity costs of the empty cells (of the optimal program) enables us to identify solutions in descending order of preference in terms of total shipping cost.

The Assignment Model

11.1 INTRODUCTION

The assignment model deals with a special class of linear-programming problems in which the objective is to assign a number of "origins" to the *same* number of "destinations" at a minimum total cost.* The assignment is to be made on a one-to-one basis. That is, each origin can associate with one and only one destination. This feature implies the existence of two specific characteristics in a linear-programming problem which, when present, give rise to an assignment problem. First, the payoff matrix for the given problem is a square matrix. Second, the optimal solution (or any solution within the given constraints) for the problem is such that there can be one and only one assignment in a given row or column of the given payoff matrix.

Payoff measures for each assignment are assumed to be known and independent of each other. With information about the number of origins and destinations and the payoff measure associated with each available assignment, the assignment model is used to choose that strategy which maximizes or minimizes the total payoff measure, depending upon whether the particular payoff represents a gain or a loss to the decision maker.

11.2 A SIMPLE ASSIGNMENT PROBLEM

Let us illustrate an extremely simple assignment problem by considering the assignment of three given jobs O_1, O_2, and O_3 to three given machines

* Of course, if the payoff measure is of the *profit* variety, the objective is to maximize total payoff.

D_1, D_2, and D_3. The problem states that any one of the jobs can be processed completely with any one of the machines. Further, the cost of processing the ith job ($i = 1, 2, 3$) with the jth machine ($j = 1, 2, 3$) is known. The objective, therefore, is to assign these jobs to the machines in a manner that will minimize the total cost of processing all the jobs. The relevant cost data are given in Table 11.1 in matrix form. A quick visual inspection of this simple problem reveals that the minimum total cost assignment will require that jobs O_1, O_2, and O_3 be assigned, respectively, to machines D_1, D_2, and D_3.

The total cost of the optimal assignment can be obtained by multiplying the cost of each assigned (occupied) cell by 1, multiplying the cost of

Table 11.1

Job	Machine		
	D_1	D_2	D_3
O_1	10	15	20
O_2	19	12	16
O_3	12	14	11

each unassigned (empty) cell by 0, and then adding the products. Thus, the total cost of the optimal assignments for this problem is

$$1(10) + 0(19) + 0(12) + 0(15) + 1(12) + 0(14)$$
$$+0(20) + 0(16) + 1(11) = \$33$$

This suggests that the optimal-assignment matrix can be represented as in Table 11.2. In other words, we can think of the assignment problem as a problem in making proper "matches" between the origins and the destinations. A value of 1 is allocated to those cells* for which a match has been made; a value of 0 is given to all other cells.

There are various methods for making these "matches." First, as will be shown in Section 11.4, we can use the transportation model for

---

* This identifies a complete utilization and satisfaction of the capacity and requirement of the particular row and column in which such a cell falls.

solving the assignment problem.   Second, provided the problem is of small dimensions, we can identify the optimal assignment by enumerating and examining all the possible assignments (see Section 11.5).   Third, we can use the "assignment model" for solving such problems (see Section 11.7).   Of these, the assignment model is the most efficient method of attack for obtaining the optimal assignment.

We have constructed and solved the problem of Table 11.1 with the objective of giving the reader an intuitive understanding of the assignment problem.   Note that the optimal-assignment matrix has one and only one

Table 11.2   *Optimal-*
              assignment
              Matrix

Job	Machine		
	D_1	D_2	D_3
O_1	1		
O_2		1	
O_3			1

assignment in each row and each column.   Another way of saying the same thing is that the sum of assignments for each row and column in the optimal solution to the assignment problem must be 1.   This requirement, along with other components of a complete statement of the assignment problem, is presented in the next section.

11.3   THE ASSIGNMENT PROBLEM AS A SPECIAL CASE OF THE TRANSPORTATION PROBLEM

It was mentioned earlier that the assignment problem is a special case of the general linear-programming problem.   As a matter of fact, the assignment problem is a special case of the "transportation problem," which, in turn, is itself a special case of the general linear-programming problem. This will become clear as we consider the 3 × 3 transportation problem in Table 11.3.   It will be recalled that a transportation problem of this type

Table 11.3

Origin	Destination			Origin capacity per time period
	D_1	D_2	D_3	
O_1	c_{11} x_{11}	c_{12} x_{12}	c_{13} x_{13}	b_1
O_2	c_{21} x_{21}	c_{22} x_{22}	c_{23} x_{23}	b_2
O_3	c_{31} x_{31}	c_{32} x_{32}	c_{33} x_{33}	b_3
Destination requirement per time period	d_1	d_2	d_3	

can be stated as:
  Minimize

$$F(X) = \sum_{j=1}^{3} \sum_{i=1}^{3} c_{ij} x_{ij}$$

subject to

$$\sum_{j=1}^{3} x_{ij} = b_i \quad i = 1, 2, 3$$

$$\sum_{i=1}^{3} x_{ij} = d_j \quad j = 1, 2, 3$$

and

$$x_{ij} \geq 0 \quad \begin{cases} i = 1, 2, 3 \\ j = 1, 2, 3 \end{cases}$$

In other words, the above transportation problem calls for the determination of the x_{ij}'s such that the objective function is minimized subject to the given constraints. Now, let us suppose that each $b_i = 1$, and each $d_j = 1$. Impose, further, the restriction that $x_{ij} = (x_{ij})^2$. Then the above transportation problem reduces to the following form:

Minimize

$$F(X) = \sum_{j=1}^{3} \sum_{i=1}^{3} c_{ij} x_{ij}$$

subject to

$$x_{ij} = (x_{ij})^2 \qquad i, j = 1, 2, 3$$

$$\sum_{j=1}^{3} x_{ij} = 1 \qquad i = 1, 2, 3 \qquad \text{struc-}$$
$$\text{tural}$$
$$\text{con-}$$
$$\sum_{i=1}^{3} x_{ij} = 1 \qquad j = 1, 2, 3 \qquad \text{straints}$$

and

$$x_{ij} \geq 0 \qquad \begin{cases} i = 1, 2, 3 & \text{nonnegativity} \\ j = 1, 2, 3 & \text{constraints} \end{cases}$$

A little reflection will show that the meaning of the structural constraints given above is as follows: (1) x_{ij} can take only the value 1 or 0 and (2) the sum of the x_{ij}'s for each row and each column is 1.   But this is exactly what was specified as one of the properties of the assignment problem in the preceding section.   In fact, the objective function, the structural constraints, and the nonnegativity constraints given above correspond exactly to the descriptive statement of the assignment problem of Table 11.1.   We therefore come to the conclusion that the assignment problem is, indeed, a special case of the transportation problem.

11.4   SOLVING AN ASSIGNMENT PROBLEM BY THE TRANSPORTATION TECHNIQUE

In so far as the assignment problem is a special case of the transportation problem, we should be able to solve any assignment problem by application of the transportation algorithm. We shall illustrate this by considering the assignment problem in Table 11.4. The number within each cell represents the cost (c_{ij}) of processing the ith job with the jth machine.

Table 11.4

Job	Machine		
	D_1	D_2	D_3
O_1	20	27	30
O_2	10	18	16
O_3	14	16	12

Table 11.5

Job	Machine			Job requirement
	D_1	D_2	D_3	
O_1	20	27	30	1
O_2	10	18	16	1
O_3	14	16	12	1
Machine capacity	1	1	1	

In view of our previous discussion, we place this problem in a transportation format in Table 11.5.

Let us use the MODI method to solve this problem. The first step, of course, is to obtain a basic feasible solution by making an initial assignment. Following the northwest-corner rule, we obtain the initial assignment given in Table 11.6a. The number of occupied cells in this assignment is 3. However, we need $m + n - 1$, that is, 5, occupied cells in order to obtain a basic feasible solution.*

We add epsilons to cells O_1D_2 and O_2D_3, thus obtaining the basic feasible solution given in Table 11.6b. An assignment of row and column numbers to this program shows that the empty cell O_2D_1 has a positive opportunity cost, since the "implied" cost of cell O_2D_1 is 11, whereas its actual cost is 10 (see Table 11.6b). Since the program of Table 11.6b does not represent an optimal assignment (the opportunity cost of cell

Table 11.6a   Initial Assignment by Northwest-corner Rule (a Degenerate Solution)

	D_1	D_2	D_3	Total
O_1	20 ①	27	30	1
O_2	10	18 ①	16	1
O_3	14	16	12 ①	1
Total	1	1	1	

Table 11.6b   A Basic Feasible Solution

	D_1	D_2	D_3	Row number
O_1	20 ①	27 ϵ	30	0
O_2	10	18 ①	16 ϵ	-9
O_3	14	16	12 ①	-13
Column number	20	27	25	

O_2D_1 is positive), we revise this program to include cell O_2D_1 (see Table 11.6d).

The revised assignment can be tested for optimality after it is made a basic feasible solution by the addition of ϵ to either cell O_2D_2 or cell O_3D_2 and a set of row and column numbers is obtained for the revised program. A visual inspection of Table 11.6e shows that the opportunity costs of all

---

* In so far as any assignment problem must have a square payoff matrix (say $n \times n$), a basic feasible solution should have $n + n - 1$ occupied cells. But, owing to the structural constraints of the assignment problem, any solution of such a problem cannot have more than n assignments (that is, n occupied cells). Hence, the assignment problem is inherently degenerate.

Table 11.6c   *First Assignment*

	D_1	D_2	D_3	Total
O_1	20 ①	27 ε	30	1
O_2	10 □	18 ①	16 ε	1
O_3	14	16 ①	12	1
Total	1	1	1	

Table 11.6d   *Revised Assignment*

	D_1	D_2	D_3	Total
O_1	20	27 ①	30	1
O_2	10 ①	18	16 ε	1
O_3	14	16	12 ①	1
Total	1	1	1	

Table 11.6e   *Revised Assignment (Optimal Assignment)*

	D_1	D_2	D_3	Row number
O_1	20	27 ①	30	0
O_2	10 ①	18 ε	16 ε	-9
O_3	14	16	12 ①	-13
Column number	19	27	25	

the empty cells are nonpositive; hence the revised assignment is the optimal assignment.

This assignment says:

Assign job O_1 to machine D_2
Assign job O_2 to machine D_1
Assign job O_3 to machine D_3

The total cost of this optimal assignment is $27 + 10 + 12 = \$49$.

By solving the assignment problem of Table 11.4 by the transportation technique, we have added little to our knowledge—since the transportation method was adequately covered in the last chapter.   The effort, however, has not been completely wasted, for we have shown that any assignment problem can be solved by the transportation technique.

11.5   SOLVING AN ASSIGNMENT PROBLEM BY ENUMERATION

The assignment problem, if time and money are assumed to be unlimited, can also be solved by first enumerating all possible assignments and then choosing the least-cost assignment.   For example, there are six possible assignments for the problem of Table 11.4.*   These assignments, along with their respective total costs, are listed in Table 11.7.   Obvi-

Table 11.7

	Assignment	Total cost, dollars
1	O_1D_1, O_2D_2, O_3D_3	$20 + 18 + 12 = 50$
2	O_1D_1, O_2D_3, O_3D_2	$20 + 16 + 16 = 52$
3	O_1D_2, O_2D_1, O_3D_3	$27 + 10 + 12 = 49$
4	O_1D_2, O_2D_3, O_3D_1	$27 + 16 + 14 = 57$
5	O_1D_3, O_2D_2, O_3D_1	$30 + 18 + 14 = 62$
6	O_1D_3, O_2D_1, O_3D_2	$30 + 10 + 16 = 56$

ously, assignment 3, with a total cost of $49, is the least-cost or optimal assignment.

Needless to say, we are not seriously advocating the solution of assignment problems by enumeration.   The reader has only to think of a 10×10 assignment problem, not to speak of larger dimensions, to realize that the solution by enumeration is impractical.†   However, this way of looking at the assignment method shows us the significance of more efficient methods of attack.

* For an assignment problem having an $n \times n$ payoff matrix, the number of possible assignments equals $n!$.   Thus, in this case we have 3!, or 6, possible assignments.

† There are 10!, or 3,628,800, possible assignments in a 10×10 assignment problem.

11.6 APPROACH OF THE ASSIGNMENT MODEL

We are now aware that it is at least theoretically possible to solve a given assignment problem by the application of the transportation technique. However, a much more efficient method of solving such problems is available. This method of solving assignment problems, known as *Flood's technique* or the *Hungarian method of assignment*, will be referred to as the *assignment model* or the *assignment method*.

The assignment method consists of three basic steps. The *first step* involves the derivation of a "total-opportunity-cost" matrix from the given payoff matrix of the problem. This is done, as we shall discuss in the next section, by (1) subtracting the lowest number of each column of the given payoff matrix from all the other numbers in its column and (2) subtracting the lowest number of each row of the matrix obtained in (1) from all the other numbers in its row. The total-opportunity-cost matrix thus derived will have at least one zero in each row and column. Any cell having an entry of zero in the total-opportunity-cost matrix is considered to be a candidate for assignment. The significance of the total-opportunity-cost matrix is that it presents some possible assignment alternatives in which the opportunity costs of some or all assignments may be zero.

The purpose of the *second step* is to determine whether an optimal assignment, guided by the total-opportunity-cost matrix derived in step 1, can be made. This is accomplished, as we shall see in the next section, by a simple test. If the test shows that an optimal assignment (with a total opportunity cost of zero) can be made, the problem is solved.* On the other hand, if an optimal assignment cannot be made, we proceed to step 3.

The purpose of the *third step* is to revise the current total-opportunity-cost matrix in order to derive some better assignments. The procedure by which this is accomplished either redistributes the zeros of the current total-opportunity-cost matrix or creates one or more new zero cells. The result is another total-opportunity-cost matrix which enables us to find a less costly assignment. In other words, the result of step 3 brings us back to the beginning of step 2, and we again search for an optimal solution. Thus, steps 2 and 3 are repeated as many times as necessary to find an optimal solution having a total opportunity cost of zero.

The remaining sections of this chapter are devoted to illustrating the development and application of the assignment method.

---

* In so far as the assignment problem involves decision making under certainty, the optimal assignment must be such that the total opportunity cost associated with the solution is zero.

11.7 DEVELOPMENT OF THE ASSIGNMENT METHOD*

In this section we shall present an intuitive rationale for the various steps of the assignment method. A brief description of the mathematical foundation of the assignment method will be given in Section 11.11.

Let us consider again the assignment problem of Table 11.4. The relevant cost data are reproduced in Table 11.8.

Table 11.8   *Original Cost Matrix*

Job	Machine		
	D_1	D_2	D_3
O_1	20	27	30
O_2	10	18	16
O_3	14	16	12

Step 1   Determining the Total-opportunity-cost Matrix

Suppose we arbitrarily choose to assign job O_3 to machine D_1. The given data show that the cost of this assignment is $14. We are aware that this is not the best assignment to machine D_1, since the assignment of job O_2 to machine D_1 would have cost us only $10 while utilizing the same full capacity of machine D_1. In other words, the decision to assign job O_3 to machine D_1 involves an opportunity cost of $4 ($14 - 10$) with respect to the lowest cost assignment in column D_1. Similarly, the decision to choose job O_1 for assignment to machine D_1 would involve an opportunity cost of $10 ($20 - 10$). Obviously, the assignment of job O_2 to machine D_1 would involve an opportunity cost of zero ($10 - 10$). In our discussion so far, we have taken a *given* machine and examined the different alternatives in terms of different jobs that can be assigned to this machine. Hence, the opportunity cost that we have been discussing so

* The discussion in this section is based on D. W. Miller and M. K. Starr, "Executive Decisions and Operations Research," chap. 10, Prentice-Hall, Inc., Englewood Cliffs, N.J., 1960.

far can be classified, in connection with our problem, as a *job-opportunity cost** or *column-opportunity cost.*

It should now be observed that we can also examine alternatives by choosing machines (rather than choosing jobs) for assignment to a given job. For example, consider again the matching of job O_3 and machine D_1. We can very well assert that since machine D_1 has been occupied by job O_3, the opportunity of assigning any other machine to job O_3 has been sacrificed. In particular, this has given rise to what may be called a *machine-opportunity cost†* or *row-opportunity cost* of \$2 (14 − 12) with respect to the lowest-cost machine.

Thus, in an assignment problem, any match between an origin and a destination gives rise to two types of opportunity costs. One is the

Table 11.9 *Job-opportunity-cost Matrix*

	D_1	D_2	D_3
O_1	10	11	18
O_2	0	2	4
O_3	4	0	0

Table 11.10 *Machine-opportunity-cost Matrix*

	D_1	D_2	D_3
O_1	0	7	10
O_2	0	8	6
O_3	2	4	0

opportunity cost with respect to the lowest payoff measure in the column to which the assignment cell belongs (we have called it job-opportunity cost). The other is the opportunity cost with respect to the lowest payoff measure in the row to which the assignment cell belongs (we have called it machine-opportunity cost). It is obvious, in view of the above discussion, that a complete job-opportunity-cost matrix (containing job-opportunity costs for each column) can be derived by simply subtracting the lowest entry in each column from all the entries in its column. For the above example, we obtain the job-opportunity-cost matrix in Table 11.9. Similarly, to arrive at a complete machine-opportunity-cost matrix, the lowest entry in each row is subtracted from all the entries in its row. Thus, for our example, we obtain the machine-opportunity-cost matrix in Table 11.10.

The total-opportunity-cost matrix can obviously be obtained by adding the job-opportunity-cost matrix to the machine-opportunity-cost matrix.

* *Ibid.,* p. 283.
† *Ibid.*

J

Thus, for our example, we have the total-opportunity-cost matrix of Table 11.11 (Table 11.9 plus Table 11.10).   The total opportunity cost, in this form, gives us only two assignments (O_2 to D_1 and O_3 to D_3) in which the total opportunity costs are zero.   But it will be recalled that we can be confident about having decided on a complete optimal solution only when the total opportunity costs of all the assignments are zero. This is possible only when each row and column has at least one zero cell. For example, if, in Table 11.11, cell O_1D_2 contained a zero, a complete optimal assignment could have been made.

It should be emphasized that total opportunity cost is a relative concept.   When we make a particular assignment which gives rise to a total opportunity cost of zero, we can be sure that this assignment is "rela-

Table 11.11   *Total-opportunity-*
cost Matrix

	D_1	D_2	D_3
O_1	10	18	28
O_2	0	10	10
O_3	6	4	0

Table 11.12   *Revised*
Total-opportunity-
cost Matrix

	D_1	D_2	D_3
O_1	0	4	18
O_2	0	6	10
O_3	6	0	0

tively" best.   Thus, we can and should manipulate the relative magnitudes of the entries in the total-opportunity-cost matrix in a search for zeros.   By subtracting the lowest entry (10) in row 1 from all the entries in row 1 of Table 11.11 and then subtracting the lowest entry in column 2 from all the entries in column 2, we obtain the total-opportunity-cost matrix in Table 11.12.   The new total-opportunity-cost matrix has four zero cells, whereas the first total-opportunity-cost matrix (Table 11.11) had two zero cells.   However, an attempt to make a complete assignment guided by the new opportunity-cost matrix will reveal that we still cannot obtain an optimal solution with a total opportunity cost of zero.*

---

* The four possible assignments which include the zero cells are:

O_1D_1, O_3D_2, O_2D_3, with a total opportunity cost of 10
O_1D_1, O_3D_3, O_2D_2, with a total opportunity cost of 6
O_2D_1, O_3D_2, O_1D_3, with a total opportunity cost of 18
O_2D_1, O_3D_3, O_1D_2, with a total opportunity cost of 4

Thus, we must obtain another total-opportunity-cost matrix, either by redistributing the current zeros or by creating one or more new zero cells. How this is accomplished is illustrated in step 3 below.

An examination of the total-opportunity-cost matrix in Table 11.12 reveals that cells O_1D_1, O_2D_1, O_3D_2, and O_3D_3 are zero cells. We could have arrived at an equivalent total-opportunity-cost matrix in which the zero cells fall at the same spots by a shorter route than that taken in the derivation of Table 11.12. The derivation of the total-opportunity-cost matrix by the shorter method involves only two sets of subtractions. First, we obtain a job-opportunity-cost matrix by subtracting the lowest entry in each column of the original cost matrix from all the entries in its

Table 11.13   *Derivation of the Total-opportunity-cost Matrix*

Original cost matrix

	D_1	D_2	D_3
O_1	20	27	30
O_2	10	18	16
O_3	14	16	12

Job-opportunity-cost matrix

	D_1	D_2	D_3
O_1	10	11	18
O_2	0	2	4
O_3	4	0	0

Total-opportunity-cost matrix

	D_1	D_2	D_3
O_1	0	1	8
O_2	0	2	4
O_3	4	0	0

column. Then we subtract the lowest entry in each row of this *derived* job-opportunity-cost matrix from all the entries in its row. The result is the total-opportunity-cost matrix in Table 11.13. Notice that the total-opportunity-cost matrix of Table 11.13 is essentially equivalent to the total-opportunity-cost matrix of Table 11.12.* In summary, then, the total-opportunity-cost matrix for any assignment problem can easily be obtained in the following manner:

1. Subtract the lowest entry in each column of the given payoff matrix from all the entries in its column.
2. Subtract the lowest entry in each row of the matrix obtained in (1) from all the numbers in its row.

* Note that the zero cells in the total-opportunity-cost matrices of Tables 11.12 and 11.13 are in exactly the same positions.

Step 2   Determining Whether an Optimal Assignment Can Be Made

An optimal assignment in this problem (guided by the total-opportunity-cost matrix of step 1) can always be made if we can locate three zero cells* in the total-opportunity-cost matrix such that a complete assignment to these cells can be made with a total opportunity cost of zero. This can happen only when no two such zero cells are in the same row or column, regardless of the number of zero cells in the total-opportunity-cost matrix. Thus, of the four zero cells in the total-opportunity-cost matrix of Table 11.13, only two zero cells are "useful" for the purpose of obtaining an optimal assignment. On the other hand, if we had an additional zero, say in cell O_1D_2, we could locate a set of three zero cells such that an optimal assignment (with a total opportunity cost of zero) could be made.

Based on the above discussion, a simple test has been devised to determine whether an optimal assignment can be made. It consists in drawing and counting the *minimum* number of horizontal and vertical lines necessary to cover *all* the zero cells in the total-opportunity-cost matrix. If the number of lines equals either the number of rows or the number of columns, an optimal assignment can be made, and the problem is solved. On the other hand, if the minimum number of lines needed to cover all the zero cells is less than the number of rows, an optimal assignment cannot be made, and it is necessary to construct a revised total-opportunity-cost matrix.

An application of the above test to the total-opportunity-cost matrix of Table 11.13 shows that an optimal assignment cannot be made at this stage. It takes only two lines (row O_3 and column D_1) to cover all the zero cells in the total-opportunity-cost matrix, whereas the number of rows is 3. Hence, a revised total-opportunity-cost matrix, which will lead us toward an optimal assignment, must be obtained.

Step 3   Revising the Current Total-opportunity-cost Matrix

After the application of the test of step 2, the cells of the total-opportunity-cost matrix can be classified into two categories: (1) the *covered cells*, which have been covered by the lines, and (2) the *uncovered cells*, which have not been covered by the lines. The presence of more than one zero cell in any covered row or column of a total-opportunity-cost

---

* For an $n \times n$ assignment problem, we must locate n such zero cells. Such zeros, no two (or more) of which lie in the same row or column, are said to form a set of *independent* zeros.

matrix may be interpreted to mean that the relative opportunity costs of some of the cells are wrong.* In order to correct this situation, we must obtain a set of three *independent* zero cells by revising the current total-opportunity-cost matrix. The result of the revision, therefore, must be such that either one of the zeros of the current total-opportunity-cost matrix (Table 11.13) is transferred to one of the uncovered nonzero cells or a new zero appears in one of the uncovered nonzero cells. Since cell O_1D_2 has the lowest opportunity cost of the uncovered nonzero cells, we would, intuitively, want this cell to emerge as a new zero cell. The procedure for accomplishing this consists in† (1) subtracting the *lowest* entry in the uncovered cells of the total-opportunity-cost matrix from all

Table 11.14a   *Current Total-opportunity-cost Matrix*

	D_1	D_2	D_3
O_1	0	1	8
O_2	0	2	4
O_3	4	0	0

→ Covering line 1

↓ Covering line 2

Table 11.14b   *Revised Total-opportunity-cost Matrix*

	D_1	D_2	D_3
O_1	0	0	7
O_2	0	1	3
O_3	5	0	0

the *uncovered* cells and (2) adding the same lowest entry to *only* those cells in which the covering lines of step 2 cross.

In our example, the lowest entry in the uncovered cells of the total-opportunity-cost matrix (Table 11.13) is 1 (cell O_1D_2). Subtracting this from all the uncovered cells, and adding it to only those cells (in this case, cell O_3D_1) in which the covering lines of step 2 cross, we obtain a revised total-opportunity-cost matrix (see Table 11.14b). An application of the test of step 2 to the revised total-opportunity-cost matrix shows that the minimum number of lines needed to cover all the zeros is 3.‡ Since the

---

* We know that the assignment problem involves decision making under certainty, and therefore there must be *at least one* strategy (assignment) involving a total opportunity cost of zero.

† The rationale for this procedure is imbedded in the mathematical theorems which will be presented in Section 11.11.

‡ Each line must be drawn in such a manner that it covers the *largest number* of zeros in the matrix.

number of rows of this matrix is also 3, an optimal assignment can be made.

Once it is established that an optimal assignment can be made, we search for a row or column in which there is only one zero cell. The first assignment is made to that zero cell, and the row and column in which this cell lies are crossed out. The remaining rows and columns of the matrix are again examined in order to find that row or column in which there remains only one zero cell. Another assignment is made, and the respective row and column are crossed out. The procedure is repeated until a complete assignment has been made. The optimal-assignment sequence for our problem is shown in Table 11.15. Since

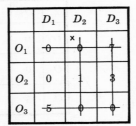

Table 11.15a *First Assignment*   **Table 11.15b** *Second Assignment*   **Table 11.15c** *Third Assignment*

cell O_3D_3 is the only zero cell in column D_3, we make the first assignment to cell O_3D_3 and cross out row O_3 and column D_3. In the reduced matrix, we note that cell O_1D_2 is the only zero cell in column D_2. Hence, we make the second assignment to cell O_1D_2 and cross out row O_1 and column D_2. This leaves only one zero cell open (cell O_2D_1), and therefore the third assignment is made to that cell. Thus, we have the following optimal assignment:

Assign job O_1 to machine D_2
Assign job O_2 to machine D_1
Assign job O_3 to machine D_3

The total opportunity cost associated with this optimal assignment is, of course, zero. The total cost of this assignment, as can be easily verified from the original cost matrix, is \$49.

11.8 PROCEDURE SUMMARY FOR THE ASSIGNMENT METHOD (MINIMIZATION CASE)

Step 1 Determine the Total-opportunity-cost Matrix

a. Arrive at a column-opportunity-cost matrix by subtracting the lowest entry of each column of the given payoff matrix from all the entries in its column.

b. Then subtract the lowest entry of each row of the matrix obtained in (*a*) from all the entries in its row.

The result of step 1*b* gives the total-opportunity-cost matrix.

Step 2 Determine Whether an Optimal Assignment Can Be Made

a. Cover *all* the zeros of the current total-opportunity-cost matrix with the *minimun possible* number of horizontal and vertical lines.

b. If the number of lines drawn in step 2*a* equals the number of rows (or columns) of the matrix, the problem can be solved. Make a complete assignment so that the total opportunity cost involved in the assignment is zero.

c. If the number of lines drawn in step 2*a* is less than the number of rows (or columns) of the matrix, proceed to step 3.

Step 3 Revise the Total-opportunity-cost Matrix

a. Subtract the lowest entry in the *uncovered* cells of the current total-opportunity-cost matrix from all the uncovered cells.

b. Add the same lowest entry to *only* those cells in which the covering lines of step 2 cross.

The result of steps 3*a* and 3*b* is a revised total-opportunity-cost matrix.

Step 4

Repeat steps 2 and 3 until an optimal assignment having a total opportunity cost of zero can be made.

11.9 THE ASSIGNMENT METHOD (MAXIMIZATION CASE)

Except for one transformation, an assignment problem in which the objective is to maximize the total payoff measure can be solved by the assignment algorithm presented above. The transformation involves subtracting all the entries of the original payoff matrix from the highest entry of the original payoff matrix. The transformed entries give us the "relative costs," and the problem then becomes a minimization problem. Once the optimal assignment for this transformed minimization problem has been identified, the total value of the original payoff measure can be found by adding the individual original entries for those cells to which the assignments have been made.

11.10 ALTERNATIVE SOLUTIONS

After having established the fact that an optimal solution for an assignment problem exists, we may find that the number and positions of the

Table 11.16 *Original Cost Matrix*

	D_1	D_2	D_3	D_4	D_5
O_1	2	4	3	5	4
O_2	7	4	6	8	4
O_3	2	9	8	10	4
O_4	8	6	12	7	4
O_5	2	8	5	8	8

zero cells in the final total-opportunity-cost matrix are such that more than one optimal assignment can be made. The presence of alternative optimal solutions for an assignment problem can be identified by the fact that, while assignments are being made via the final* total-opportunity-cost matrix, a row or column containing only one zero cell cannot be located. Let us illustrate by considering the assignment problem in Table 11.16.

By subtracting the lowest entry in each column from all the entries in

---

* Such as the revised total-opportunity-cost matrix of Table 11.14b. It is "final" in the sense that we know that this matrix can guide us in making an optimal assignment, and thus it need not be revised.

its column (step 1*a*), we obtain the job-opportunity-cost matrix (see Table 11.17).   In this particular problem, the job-opportunity-cost matrix is also the total-opportunity-cost matrix.   When we subtracted the lowest entry in each column from all the entries in its column, we obtained a zero cell in each row of the matrix.   Thus, if we now subtract the lowest entry in each row from all the entries in its row (step 1*b*), the matrix will not be changed.   Hence, the matrix of Table 11.17 is indeed the total-opportunity-cost matrix.

Next, we must draw the *minimum* number of horizontal and vertical lines needed to cover all the zeros of the total-opportunity-cost matrix (step 2).   The particular sequence in which these lines must be drawn in

Table 11.17

	D_1	D_2	D_3	D_4	D_5
O_1	0	0	0	0	0
O_2	5	0	3	3	0
O_3	0	5	5	5	0
O_4	6	2	9	2	0
O_5	0	4	2	3	4

our example is shown in Table 11.18.*   Since we need only four lines to cover all the zeros of the total-opportunity-cost matrix, whereas the number of rows in the matrix is 5, an optimal assignment cannot be made at this stage.

The next step is to revise this total-opportunity-cost matrix by the application of step 3 of the assignment algorithm.   An examination of the uncovered cells of the current total-opportunity-cost matrix (Table 11.18) indicates that the smallest entry in an uncovered cell is 2.   We therefore subtract 2 from all the uncovered cells and add 2 to only those cells in which the covering lines cross (O_1D_1, O_1D_5, O_2O_1, and O_2O_5).   Thus, the revised total-opportunity-cost matrix in Table 11.19 is obtained. An attempt to draw lines to cover the zeros of the matrix shows that a

* Line 4 could have been drawn to cover column D_2 rather than row O_2.   This, however, would have made no difference in the result of this step.

minimum number of 5 lines is required. The number of rows in the matrix is also 5; so we conclude that an optimal assignment can now be made.

Table 11.18 *Application of Step 2*

	D_1	D_2	D_3	D_4	D_5	
O_1	0	0	0	0	0	→ Covering line 1
O_2	5	0	3	3	0	→ Covering line 4
O_3	0	5	5	5	0	
O_4	6	2	9	2	0	
O_5	0	4	2	3	4	

Covering line 3   Covering line 2

Table 11.19 *Revised Total-opportunity-cost Matrix*

	D_1	D_2	D_3	D_4	D_5
O_1	2	0	0	0	2
O_2	7	0	3	3	2
O_3	0	3	3	3	0
O_4	6	0	7	0	0
O_5	0	2	0	1	4

An examination of the rows and columns of the matrix in Table 11.19 shows that there is only one row (row O_2) containing a single zero cell. Hence, the first assignment is made to this cell (O_2D_2). This means that row O_2 and column D_2 can be deleted from the matrix, and we are left with the matrix of Table 11.20. An examination of the reduced matrix shows that there is no row or column in which there is only one zero cell. This fact, as we mentioned earlier, indicates the presence of alternative solutions in an assignment problem.

Let us arbitrarily make the second assignment to cell O_1D_3.   Then the remaining three assignments may be made in this order: third assignment to cell O_5D_1, fourth assignment to cell O_4D_4, and fifth assignment to cell O_3D_5.   On the other hand, if the second assignment had been made to, say, cell O_3D_1, the remaining three assignments would have been to cells O_5D_3, O_4D_5, and O_1D_4.   The total costs of these alternative optimal

Table 11.20

	D_1	D_3	D_4	D_5
O_1	2	0	0	2
O_3	0	3	3	0
O_4	6	7	0	0
O_5	0	0	1	4

assignments are, of course, the same ($20).   The two alternative optimal assignments are given below:

First optimal assignment:

Assign O_2 to D_2
Assign O_1 to D_3
Assign O_5 to D_1
Assign O_4 to D_4
Assign O_3 to D_5

Second optimal assignment:

Assign O_2 to D_2
Assign O_3 to D_1
Assign O_5 to D_3
Assign O_4 to D_5
Assign O_1 to D_4

A little reflection will show that, for the assignment problem under consideration, there are many more alternative solutions.

11.11   MATHEMATICAL FOUNDATION OF THE ASSIGNMENT ALGORITHM

The foundation of the assignment algorithm is provided by (1) a mathematical theorem proved by the Hungarian mathematician Konig and (2) a statement of a certain matrix property.*

The theorem states: *If the elements of a matrix are divided into two classes by a property R, then the minimum number of lines that contain all the elements with the property R is equal to the maximum number of elements with the property R, with no two on the same line.*

The particular matrix property is: *Given a cost matrix $A = [a_{ij}]$, if we form another matrix $B = [b_{ij}]$, where $b_{ij} = a_{ij} - u_i - v_j$, and where u_i and v_j are arbitrary constants, the solution of A is identical with that of B.*

In order to clarify the relationship of the solution stages, based on the above-mentioned theorem and matrix property, to the corresponding steps of the assignment algorithm, we shall solve the problem of Table 11.4 by these two approaches.

Solution by the Assignment Algorithm

Original cost matrix

	D_1	D_2	D_3
O_1	20	27	30
O_2	10	18	16
O_3	14	16	12

Solution Based on the Konig Theorem and the Matrix Property

Original cost matrix

$A = [a_{ij}] =$

	D_1	D_2	D_3
O_1	20	27	30
O_2	10	18	16
O_3	14	16	12

---

* The mathematical theorem, the matrix property, and references are given in C. W. Churchman, R. L. Ackoff, and E. L. Arnoff, "Introduction to Operations Research," chap. 12, John Wiley & Sons, Inc., New York, 1957.

Step 1

a. Subtract the lowest entry in each column of the original cost matrix from all the entries in its column. The above matrix becomes

	D_1	D_2	D_3
O_1	10	11	18
O_2	0	2	4
O_3	4	0	0

Step 1

a. Transform the given payoff matrix by the relationship $b_{ij} = a_{ij} - u_i - v_j$. In particular, let $u_1^{(0)} = u_2^{(0)} = u_3^{(0)} = 0$, and let $v_1^{(0)} = 10$, $v_2^{(0)} = 16$, and $v_3^{(0)} = 12.$* Then,

	D_1	D_2	D_3	$u_i^{(0)}$
O_1	20	27	30	0
O_2	10	18	16	0
O_3	14	16	12	0
$v_j^{(0)}$	10	16	12	

$$\Rightarrow [b_{ij}] = \begin{array}{|c|c|c|} \hline 10 & 11 & 18 \\ \hline 0 & 2 & 4 \\ \hline 4 & 0 & 0 \\ \hline \end{array}$$

b. Subtract the lowest entry in each row of the matrix obtained in step 1*a* from all the entries in its row. The above matrix becomes

b. Transform the matrix derived above by the relationship $b_{ij}' = b_{ij} - u_i - v_j$. In particular, let $u_1^{(1)} = 10$, $u_2^{(1)} = 0$, $u_3^{(1)} = 0$, and $v_1^{(1)} = v_2^{(1)} = v_3^{(1)} = 0$.

---

* u_i's refer to the row numbers, and v_j's refer to the column numbers. Furthermore, the subscripts refer to the particular row or column, whereas the superscripts refer to the sequence of transformations. For example, $u_1^{(0)}$ is the value to be subtracted from all the entries of the first row of the original matrix. Similarly, $v_1^{(1)}$ would be the value to be subtracted from all the entries of the first column of the matrix derived after the first iteration, and so on.

Then,

	D_1	D_2	D_3
O_1	0	1	8
O_2	0	2	4
O_3	4	0	0

	D_1	D_2	D_3	$u_i^{(1)}$
O_1	10	11	18	10
O_2	0	2	4	0
O_3	4	0	0	0
$v_j^{(1)}$	0	0	0	

$$\Rightarrow [b'_{ij}] = \begin{array}{|c|c|c|} \hline 0 & 1 & 8 \\ \hline 0 & 2 & 4 \\ \hline 4 & 0 & 0 \\ \hline \end{array}$$

Step 2

Cover all the zeros of the total-opportunity-cost matrix with the minimum possible number of horizontal and vertical lines.

Since only two lines are needed to cover all the zeros, whereas the number of rows of the matrix is 3, the optimal assignment cannot be made at this stage.

Proceed to step 3.

Step 2

An application of the Konig theorem to this matrix $[b'_{ij}]$ shows that, since the minimum number of lines to cover the zero cells is 2, the maximum number of "real" zero cells is also 2. However, we need three real zero cells to make an optimal assignment. Hence, the above matrix must be further transformed. Note that the so-called "property R" in the theorem corresponds to a total opportunity cost of zero in a given cell.

Step 3

Subtract the lowest entry in the uncovered cells from all the uncovered cells, and add it to only those

Step 3

Transform the previous matrix $B' = [b'_{ij}]$ by the relationship $b''_{ij} = b'_{ij} - u_i - v_j$. In particular, let

cells in which the covering lines of step 2 cross. The matrix of step 1b becomes

	D_1	D_2	D_3
O_1	0	0	7
O_2	0	1	3
O_3	5	0	0

$u_1^{(2)} = 1$, $u_2^{(2)} = 1$, $u_3^{(2)} = 0$, and let $v_1^{(2)} = -1$, $v_2^{(2)} = 0$, and $v_3^{(2)} = 0$. Then,

	D_1	D_2	D_3	$u_i^{(2)}$
O_1	0	1	8	1
O_2	0	2	4	1
O_3	4	0	0	0
$v_j^{(2)}$	-1	0	0	

$$\Rightarrow [b_{ij}''] = \begin{array}{|c|c|c|} \hline 0 & 0 & 7 \\ \hline 0 & 1 & 3 \\ \hline 5 & 0 & 0 \\ \hline \end{array}$$

Step 4

Since it takes a minimum of three lines to cover all the zero cells of the above total-opportunity-cost matrix and the number of rows is also 3, an optimal assignment having a total opportunity cost of zero can be made. The optimal assignment is O_2 to D_1, O_1 to D_2, and O_3 to D_3.

Step 4

An application of the Konig theorem to this matrix $[b_{ij}'']$ shows that, since the minimum number of lines needed to cover the zero cells is 3, the maximum number of real zero cells is also 3. Hence, an optimal assignment can be made. The optimal assignment is O_2 to D_1, O_1 to D_2, and O_3 to D_3.

Step-by-step comparison of the two approaches for solving an assignment problem shows that the assignment algorithm, in essence, repeatedly applies the Konig theorem to the given payoff matrix until a set of independent zeros is discovered. It is hoped that the comparison of the two approaches will give the reader a thorough understanding of the assignment algorithm.

It may be of interest to point out that the u_i's and v_j's of the mathematical theorem are the dual variables of the corresponding primal assignment problem. They also correspond to the row and column numbers of the transportation technique.

The Meaning of Linearity

In order to grasp the nature of linear-programming problems, one must be familiar with such terms as *variables, functions,* and *linear equations.* An attempt will be made in the following paragraphs to give some "tangible" meaning to these terms.

I.1 CONSTANTS, DEPENDENT VARIABLES, INDEPENDENT VARIABLES

Suppose that an industrial worker has a wage rate of $2 per hour. Then, if he works 40 hours in a particular week, his total earnings for that week are $80 (2 × 40). During some other week he may have worked for a total of only 30 hours, in which case he would have earned $60. In any case, his total earnings for a particular week, assuming no overtime, can always be calculated as follows:

Total earnings = 2 × number of hours worked

If we let

h = number of hours worked
T = total earnings

then

$$T = 2h$$

is the relationship between this worker's total earnings and the number of hours worked.

Similarly, another worker might have a wage rate of $4 per hour; for him, the equation $T = 4h$ would hold. Briefly, if the number of hours worked were the only criterion for wage payment, we could say that $T = Kh$, where K is a constant, for a particular worker or for a particular class of workers, to be determined in a specific situation.

In this simple illustration we note two kinds of mathematical quantities. One type (the quantity 2 in this example) remains *fixed*. The other type, exemplified by T and h, is allowed to *vary*. The quantity K which remains fixed for a *given* problem is called a *constant*. Since such constants are arbitrarily assigned or determined to reflect the situation at hand and can change in value from problem to problem, they are sometimes called *arbitrary constants* or *parameters*. On the other hand, mathematical constants such as e or π, since they assume the same value from problem to problem, are called *absolute constants*.

Let us now examine the behavior of T and h. First, it is obvious that the value of T is determined by choosing a value for h. Second, both h and T can vary. Quantities such as h and T, since they can assume various values in a given problem, are called *variables*. Further, since the value of T is dependent on h and is determined by assigning a value to h, T is called a *dependent variable*, while h is called an *independent variable*.

I.2   THE CONCEPT OF A FUNCTION

A function is a rule which relates different variables. The fact that T above was determined by h can also be represented by the notation $T = f(h)$, which reads, "Total earnings are a *function* of hours worked." It does *not* mean that T is equal to f times h.

The total earnings of our industrial worker were determined by only one independent variable, namely, h. In some cases, there may be more than one independent variable determining the value of the dependent variable. For example, for a company producing only one type of product, profit is determined by taking the difference between unit revenue and unit cost for that product and multiplying it by the number of units sold. This general fact can be stated with the following notation:

$$p = f(n, r, c)$$

where p = total profit, dollars
   n = number of units sold, physical units
   r = revenue per unit, dollars
   c = cost per unit, dollars
The above equation reads: "Total profit is a function of n, r, and c."

This functional notation is meant to give the "generalized" idea that certain variables are somehow related.   An explicit statement of equality among the variables will "particularize" the relationship.   Thus, in our earlier example, the equation $T = 2h$ particularized the functional relationship $T = f(h)$.   In other words, *a rule of correspondence* between h and T was given.   Thus a function is sometimes defined as a rule of correspondence among variables.

As mentioned earlier, once a function is particularized it establishes an equality between different quantities on opposite sides of an equal-sign. Therefore, an equation is essentially a statement of equality between different quantities.   But an *equation* is true only for certain values of the independent variables.   For example, the equation $y - 2 = 9$ is true only if the variable y assumes a value of 11.   On the other hand, the equation $4(x + y) = 4x + 4y$, involving the two variables x and y, holds for all possible values of x and y.   An equation which is true for all possible values of the variables is called an *identity*.

I.3   DOMAIN AND RANGE

Let us go back to our original example.   The worker may be employed in a factory in which overtime work is never permitted.   Obviously, his weekly working hours can vary only from a minimum of zero hours to a maximum of 40 hours.   In other words, h can only assume one of the following values:

$$h = 0, 1, 2, \ldots, 40$$

The possible values that the independent variable h can assume are said to comprise the *range* of the independent variable and the *domain* of the function.

Corresponding to each of the possible values of h, the dependent variable T takes the following values:

$$T = 0, 2, 4, \ldots, 80$$

The possible values that the dependent variable T may take are said to comprise the *range* of the dependent variable or the *range* of the function.

I.4   SINGLE-VALUED VERSUS MULTIPLE-VALUED FUNCTIONS

In our example, for every possible value of the independent variable h, the dependent variable T obviously takes one and only one value. A function such as this ($T = Kh$) is said to be a *single-valued* function of h. On the other hand, in the equation of a circle with center at the origin ($x^2 + y^2 = r^2$), for a given radius r (say 2 feet), we note a functional relation of the form $y = f(x, r)$ which is not single-valued. In particular, here,

$$y^2 = r^2 - x^2 = 4 - x^2$$

or

$$y = \pm \sqrt{4 - x^2}$$

In other words, for a particular value of x, the variable y assumes two values. This type of function, then, is a *multiple-valued* function of x. When working with multiple-valued functions, the analyst must choose only the solution(s) applicable to the problem at hand. In linear-programming problems we deal with single-valued linear functions only.

I.5   LINEAR EQUATIONS

A *linear* equation is a relationship between quantities in which, when reduced to its simplest form, the sum of the exponents of the variables in any one term adds up to 1. For example, $Y = 4 + 2X$ is a linear equation relating the variables X and Y. When we say that the relationship between, say, two variables X and Y is linear, we mean that additions of the same magnitude to the one have a constant effect on the other. If we assume X to be the independent and Y the dependent variable, the value of Y is determined by the value assigned to or taken by X. Thus, we say that Y is a function of X. The functional statement represented by the notation $Y = f(X)$ states that Y is a function of X but does not give any information about the exact relationship between X and Y. On the other hand, $Y = f(X) = 4 + 2X$, which is a linear relationship, is a complete statement of the relationship between X and Y.

Further discussion of linear equations is presented in the following examples.

Example 1

The equation $X = 5$ is a linear equation. Notice that X is the only variable in this equation, and its exponent is 1. The relationship

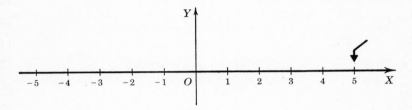

Figure I.1   Representation of the linear equation $X = 5$ as a point.

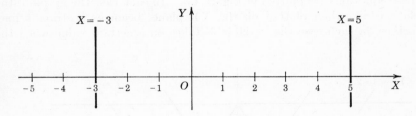

Figure I.2   Representation of the linear equations $X = 5$ and $X = -3$ as lines.

$X = 5$, if considered in only *one direction*, represents a point on the X axis of the traditional XY plane (see Figure I.1). This point can be plotted by moving 5 units to the right of the origin O. The same equation ($X = 5$), if considered in a two-dimensional space, becomes a straight line. This straight line is obtained by first plotting a point which is 5 units to the right of the origin O and then drawing a line through this point parallel to the Y axis (see Figure I.2). Similarly, $X = -3$ is another linear equation which can be plotted either as a point, by moving 3 units to the left of the origin O, or as a line parallel to the Y axis passing through the point $X = -3$.

Example 2

Now consider a linear equation of the form $Y = a + bX$, which is a relationship in two dimensions—horizontal and vertical. Traditionally, Y is plotted in the vertical direction, and X, as before, is plotted in the horizontal direction (see Figure I.3). In this case, a and b are constants whose values are to be assigned. Let us examine the following equations of the general form $Y = a + bX$:

$$Y = 4 + 2X \tag{1}$$

$$Y = -4 + 2X \tag{2}$$

$$Y = 4 - 2X \tag{3}$$

$$Y = -4 - 2X \tag{4}$$

$$Y = 2X \tag{5}$$

$$Y = -2X \tag{6}$$

$$Y = 4 \tag{7}$$

$$Y = -4 \tag{8}$$

These equations are plotted in Figure I.3. In each case the graph of the relationship, when plotted on the XY plane, becomes a straight line. Further, in each case the variable Y takes on a certain value when the

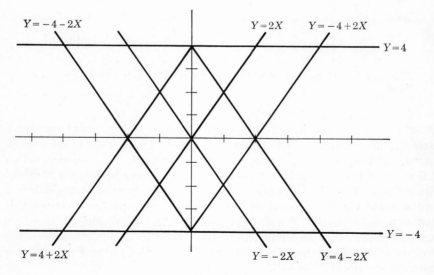

Figure I.3

variable X is zero. In Equation (1), for example, when $X = 0$, Y equals 4. In Equation (2), on the other hand, when $X = 0$, $Y = -4$. Vertical distances such as 4 and -4, obtained by letting X equal zero in equations of the form $Y = a + bX$, are called Y *intercepts*. The coefficient of X in Equations (1) and (2) is the quantity $+2$. This coefficient is called the *slope* of the equation. Briefly, the slope represents a rate. Here, in Equation (1), the slope $+2$ means that, for every unit of change in the variable X, the variable Y changes by 2 units in the *same* direction.

In Equation (3), on the other hand, since the slope is -2, the variable Y changes by 2 units for every unit of change in the variable X, but its direction of change is *opposite* the direction of change of X. Thus, if the coefficient of X in an equation of the form $Y = a + bX$ is a positive

Table I.1

X	$Y = 4 + 2X$	$Y = -4 + 2X$	$Y = 4 - 2X$	$Y = -4 - 2X$	$Y = 2X$	$Y = -2X$
0	4	-4	4	-4	0	0
1	6	-2	2	-6	2	-2
2	8	0	0	-8	4	-4
3	10	2	-2	-10	6	-6
4	12	4	-4	-12	8	-8

quantity, the equation has a positive slope, and changes in the variables take place in the same direction. By the same token, a negative coefficient of X indicates a negative slope; changes in the variables take place in opposite directions. Table I.1 gives the changes in the variable Y corresponding to changes in the variable X for Equations (1) through (6).

The relationship $Y = a + bX$ is called the *slope-intercept form* of a linear equation; a is the Y intercept, and b is the slope.

Example 3

Other types of linear equations are those in which more than two variables are related. For example, assume the data in Table I.2 for a manufacturing department producing three products X, Y, and Z. The problem here may be to choose a particular combination of the products X, Y, and Z to maximize profit contribution. However, assuming that the optimal combination of products is such that the available capacities of machines 1 and 2 will be fully utilized, we can write the following equations:

For machine 1:

$$2X + 2Y + 3Z = 100$$

For machine 2:

$$3X + 3Y + 4Z = 120$$

Each of the above equations is a three-dimensional linear equation. Any three-dimensional linear equation, if graphed, will yield a plane. Notice

Table I.2

Machine	Hours required to produce product			Machine capacity per time period, hours
	X	Y	Z	
1	2	2	3	100
2	3	3	4	120
Profit contribution per unit.......	$10	$15	$20	

that in both the above equations the sum of the exponents of the variables in any one term is 1. So far we have discussed linear relationships in spaces involving not more than three dimensions. The idea can of course be extended to more than three dimensions, although actual graphing is not possible in such cases. For example, a linear equation involving more than three variables will give rise to what we call a *hyperplane*.

I.6   TESTS FOR LINEARITY OF A FUNCTION

In linear-programming problems the objective function must be a linear function. In general, any first-degree polynomial such as

$$f(X) = a_1X_1 + a_2X_2 + \cdots + a_nX_n$$

is a linear function involving n dimensions. The mathematical test for the linearity of a given function $f(X)$ is the satisfaction of the following two conditions:*

* John G. Kemeny, J. Laurie Snell, and G. L. Thomson, "Finite Mathematics," p. 233, Prentice-Hall, Inc., Englewood Cliffs, N.J., 1957.

Condition 1

Multiplying each variable in the linear function by a constant k gives the same result as multiplying the functional value by k:

$$f(kX) = kf(X)$$

Suppose that $f(X)$ represents a profit function of the form

$$\text{Profit} = f(X) = 10X_1 + 15X_2 + 20X_3$$

where X_1, X_2, and X_3 represent a given product mix.   Then condition 1 says that a k-fold increase in the current output (in exactly the same proportion) will result in increasing the current profit k times.

Condition 2

If there are two separate programs for a given problem whose profit functions are represented by $f(X_1)$ and $f(X_2)$, then these values are additive:

$$f(X_1) + f(X_2) = f(X_1 + X_2)$$

The above condition simply means that two separate program structures, considered separately or together, will have additive effects on the objective function.

We complete the discussion on linearity by quoting Forrester:*

A linear model is one in which the concept of "superposition" holds.   In a linear system the response to every disturbance runs its course independently of preceding or succeeding inputs to the system; the total result is no more nor less than the sum of the separate components of system response.   The response to an input is independent of when the input occurs in the case of a linear system with constant co-efficients (not for a linear system having time-varying co-efficients).   Only damped or sustained oscillations can exist in an actual linear system; an oscillation that grows is not bounded and must become explosively larger.   These are not descriptions of real industrial and economic systems. . . .

* J. W. Forrester, "Industrial Dynamics," p. 50, The M.I.T. Press, Cambridge, Mass., and John Wiley & Sons, Inc., New York, 1961.

A Note on Inequalities

The technical specifications of linear-programming problems can usually be translated into linear inequalities. For example, consider the data in Table II.1 for a linear-programming problem. The objective function

Table II.1

Machine	Hours required to produce product			Machine capacity per time period, hours
	X	Y	Z	
1	2	2	4	100
2	3	4	6	120
Profit contribution per unit.......	$10	$15	$20	

to be maximized in this problem is obviously

$$10X + 15Y + 20Z$$

Relating the technical specifications of the problem to the available capacities of the resources, we can write the following equations:

For machine 1:

$$2X + 2Y + 4Z = 100 \tag{1}$$

250

For machine 2:

$$3X + 4Y + 6Z = 120 \tag{2}$$

Equations (1) and (2) will hold only if a program can be designed such that the available capacities of the two machines are fully utilized. In other words, products X, Y, and Z must be produced in such quantities that *exactly* 100 hours on machine 1 and *exactly* 120 hours on machine 2 are utilized. This complete utilization of available resources may not always be possible, for our objective of maximizing profit contribution, in this case, may dictate the manufacture of such quantities of the products X, Y, and Z that a portion of the available resources need not be utilized. Of course we can never utilize more than 100 hours on machine 1 and 120 hours on machine 2, since these numbers represent the upper limits of the available capacities of the two machines.

In view of the above discussion, a more realistic way of stating this problem would be to say that X, Y, and Z must be produced in such quantities "that machine utilization is less than or equal to the available capacities." Thus, the technical specifications of this problem can be translated into the *statements of inequalities* given below:

For machine 1:

$$2X + 2Y + 4Z \leq 100$$

For machine 2:

$$3X + 4Y + 6Z \leq 120$$

The symbol $\leq$ is used to denote an inequality of the "less than or equal to" type.

In another situation, a person may stipulate that one of his requirements for accepting a new job is that the salary be greater than or equal to $8,000 per year. This is a statement of inequality which can be represented as follows:

Required salary $\geq$ $8,000

In particular, the following inequality signs are used in formulating relationships between different quantities:

Symbol	*Reads*
>	Greater than
≥	Greater than or equal to
<	Less than
≤	Less than or equal to

The rules for manipulating equalities and those for inequalities are the same, except for the following exceptions, which hold only for inequalities:

1. If both sides of an inequality are multiplied by the same negative quantity, the inequality sign changes direction.

   Example

   $$-4 < -3$$

   But

   $$(-1)(-4) > (-1)(-3)$$

   That is,

   $$4 > 3$$

2. If both sides of an inequality are divided by the same negative quantity, the inequality sign changes direction.

   Example

   $$-4 < -3$$

   But

   $$\frac{-4}{-1} > \frac{-3}{-1}$$

   or

   $$4 > 3$$

As we have seen, a realistic view of the data in Table II.1 necessitated the use of inequalities rather than equations to represent the problem specifications. Similarly, inequalities are often employed to represent business situations, and, in general, restrictions of various sorts are usually expressed by inequalities.

A System of Linear Equations Having a Unique Solution

Solve for X_1, X_2, and X_3 in the following set of linear equations:

$$\begin{aligned}
X_1 \qquad + 2X_3 &= 4 \\
X_1 + X_2 \qquad &= 5 \\
4X_1 + 3X_2 + X_3 &= 2
\end{aligned} \tag{1}$$

The above equations can be transformed into matrix notation as follows:

$$AX = P \tag{2}$$

where

$$A = \begin{bmatrix} 1 & 0 & 2 \\ 1 & 1 & 0 \\ 4 & 3 & 1 \end{bmatrix} \qquad X = \begin{bmatrix} X_1 \\ X_2 \\ X_3 \end{bmatrix} \qquad P = \begin{bmatrix} 4 \\ 5 \\ 2 \end{bmatrix}$$

The determinant of A is

$$\begin{aligned}
|A| &= \begin{vmatrix} 1 & 0 & 2 \\ 1 & 1 & 0 \\ 4 & 3 & 1 \end{vmatrix} = a_{11}C_{11} + a_{12}C_{12} + a_{13}C_{13} \\
&= 1 \begin{vmatrix} 1 & 0 \\ 3 & 1 \end{vmatrix} + 0C_{12} + 2 \begin{vmatrix} 1 & 1 \\ 4 & 3 \end{vmatrix} \\
&= 1 + 0 - 2 = -1
\end{aligned}$$

Since the determinant has a nonzero value, a unique solution exists (see Figure III.1). To obtain this solution, premultiply (2) by the inverse of

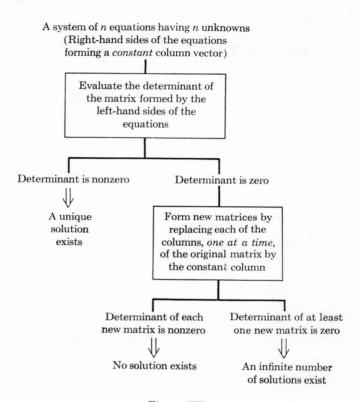

Figure III.1

A. Then

$$A^{-1}AX = A^{-1}P$$

or

$$IX = A^{-1}P$$

or

$$X = A^{-1}P$$

But

$$A^{-1} = \frac{A_{adj}}{|A|}$$

where

$$A_{adj} = \begin{bmatrix} C_{11} & C_{21} & C_{31} \\ C_{12} & C_{22} & C_{32} \\ C_{13} & C_{23} & C_{33} \end{bmatrix} = \begin{bmatrix} 1 & 6 & -2 \\ -1 & -7 & 2 \\ -1 & -3 & 1 \end{bmatrix}$$

$$\therefore \quad A^{-1} = \frac{A_{adj}}{|A|} = \frac{A_{adj}}{-1} = \begin{bmatrix} -1 & -6 & 2 \\ 1 & 7 & -2 \\ 1 & 3 & -1 \end{bmatrix}$$

Hence,

$$X = A^{-1}P = \begin{bmatrix} -1 & -6 & 2 \\ 1 & 7 & -2 \\ 1 & 3 & -1 \end{bmatrix} \begin{bmatrix} 4 \\ 5 \\ 2 \end{bmatrix} = \begin{bmatrix} -30 \\ 35 \\ 17 \end{bmatrix}$$

$$X = \begin{bmatrix} X_1 \\ X_2 \\ X_3 \end{bmatrix} = \begin{bmatrix} -30 \\ 35 \\ 17 \end{bmatrix}$$

Thus, $X_1 = -30$, $X_2 = 35$, and $X_3 = 17$.   These values of X_1, X_2, and X_3 represent a unique solution to the original system of linear equations.

A System of Linear Equations Having No Solution

Solve for X_1, X_2, and X_3 in the following set of linear equations:

$$2X_1 \qquad + 2X_3 = 4 \tag{1}$$

$$X_1 + X_2 \qquad = 5 \tag{2}$$

$$4X_1 + 3X_2 + X_3 = 2 \tag{3}$$

The above equations can be transformed into matrix notation as follows:

$$AX = P \tag{4}$$

where

$$A = \begin{bmatrix} 2 & 0 & 2 \\ 1 & 1 & 0 \\ 4 & 3 & 1 \end{bmatrix} \qquad X = \begin{bmatrix} X_1 \\ X_2 \\ X_3 \end{bmatrix} \qquad P = \begin{bmatrix} 4 \\ 5 \\ 2 \end{bmatrix}$$

The determinant of A is

$$|A| = \begin{vmatrix} 2 & 0 & 2 \\ 1 & 1 & 0 \\ 4 & 3 & 1 \end{vmatrix} = a_{11}C_{11} + a_{12}C_{12} + a_{13}C_{13}$$

$$= 2\begin{vmatrix} 1 & 0 \\ 3 & 1 \end{vmatrix} + 0C_{12} + 2\begin{vmatrix} 1 & 1 \\ 4 & 3 \end{vmatrix}$$

$$= 2(1) + 0 + 2(3 - 4) = 0$$

Since the determinant is zero, no unique solution exists. In order to determine whether the system has no solution or an infinite number of solutions, we form a new matrix A' by replacing the third column of A with the constant column P.*   Thus,

$$A' = \begin{bmatrix} 2 & 0 & 4 \\ 1 & 1 & 5 \\ 4 & 3 & 2 \end{bmatrix}$$

The determinant of A' is

$$|A'| = \begin{vmatrix} 2 & 0 & 4 \\ 1 & 1 & 5 \\ 4 & 3 & 2 \end{vmatrix} = a_{11}C_{11} + a_{12}C_{12} + a_{13}C_{13}$$

$$= 2\begin{vmatrix} 1 & 5 \\ 3 & 2 \end{vmatrix} + 0C_{12} + 4\begin{vmatrix} 1 & 1 \\ 4 & 3 \end{vmatrix}$$

$$= -26 - 4 = -30$$

Since the determinant of the new matrix A' is nonzero, the original system of equations has no solution.

To illustrate this, we first solve for X_1 and X_2 from (1) and (2), in terms of X_3.   From Equation (1),

$$X_1 = \tfrac{1}{2}(4 - 2X_3) = 2 - X_3 \tag{5}$$

Substituting (5) in Equation (2),

$$X_2 = 5 - X_1 = 5 - (2 - X_3) = 3 + X_3 \tag{6}$$

Now, to solve for X_1 and X_2, in terms of X_3, from Equations (2) and (3), we proceed as follows: Multiplying Equation (2) by 3, we obtain

$$3X_1 + 3X_2 = 15 \tag{7}$$

Subtracting (7) from (3),

$$X_1 = -13 - X_3 \tag{8}$$

* See Figure III.1.

Substituting (8) in (2),

$$X_2 = 5 - X_1 = 5 - (-13 - X_3) = 18 + X_3 \tag{9}$$

Clearly, the values of X_1 and X_2 in (5) and (6) do not agree with the values of X_1 and X_2 in (8) and (9). Hence, the original system of equations is inconsistent and has no solution.

A System of Linear Equations Having an Infinite Number of Solutions

Solve for X_1, X_2, and X_3 in the following set of linear equations:

$$2X_1 \qquad + 2X_3 = 4 \tag{1}$$

$$X_1 + X_2 \qquad = 2 \tag{2}$$

$$4X_1 + 3X_2 + X_3 = 8 \tag{3}$$

The above equations can be transformed into matrix notation as follows:

$$AX = P \tag{4}$$

where

$$A = \begin{bmatrix} 2 & 0 & 2 \\ 1 & 1 & 0 \\ 4 & 3 & 1 \end{bmatrix} \qquad X = \begin{bmatrix} X_1 \\ X_2 \\ X_3 \end{bmatrix} \qquad P = \begin{bmatrix} 4 \\ 2 \\ 8 \end{bmatrix}$$

The determinant of A is

$$|A| = \begin{vmatrix} 2 & 0 & 2 \\ 1 & 1 & 0 \\ 4 & 3 & 1 \end{vmatrix} = 0$$

Since the determinant is zero, no unique solution exists. In order to determine whether the system has no solution or an infinite number of solutions, we form a new matrix A' by replacing the first column of A with the constant column P.* Thus,

$$A' = \begin{bmatrix} 4 & 0 & 2 \\ 2 & 1 & 0 \\ 8 & 3 & 1 \end{bmatrix}$$

The determinant of A' is

$$|A'| = \begin{vmatrix} 4 & 0 & 2 \\ 2 & 1 & 0 \\ 8 & 3 & 1 \end{vmatrix} = a_{11}C_{11} + a_{12}C_{12} + a_{13}C_{13}$$

$$= 4 \begin{vmatrix} 1 & 0 \\ 3 & 1 \end{vmatrix} + 0C_{12} + 2 \begin{vmatrix} 2 & 1 \\ 8 & 3 \end{vmatrix}$$

$$= 4(1) + 0 + 2(6 - 8) = 4 - 4 = 0$$

Similarly, the reader can verify that the determinants of the two matrices formed by replacing the second and the third columns of A by the constant column P will also be zero. Since the determinant of each new matrix A' is also zero, the original system of equations has an infinite number of solutions.

To illustrate this, we examine the original matrix A for some 2×2 submatrix whose determinant is nonzero.† Any 2×2 submatrix in A has a nonzero determinant. Let us consider the submatrix $\begin{bmatrix} 2 & 0 \\ 1 & 1 \end{bmatrix}$ which is associated with X_1 and X_2 in Equations (1) and (2). We treat X_3 as a constant, transfer it to the right-hand side of Equation (1), and examine the following version of Equations (1) and (2):

$$2X_1 \qquad = 4 - 2X_3 \qquad\qquad\qquad (5)$$

$$X_1 + X_2 = 2 \qquad\qquad\qquad (6)$$

From (5),

$$X_1 = 2 - X_3 \qquad\qquad\qquad (7)$$

* See Figure III.1.
† Having established that $|A| = 0$, we know that at most $n - 1$, that is, $3 - 1$, of the original set of vectors can be linearly independent. Thus we examine all 2×2 submatrices.

Substituting (7) in (6),

$$X_2 = 2 - X_1 = 2 - (2 - X_3) = X_3 \qquad (8)$$

Now, since X_3 is treated as a constant, it can equal any number, say C. Then

$$X_3 = C$$
$$X_2 = X_3 = C$$
$$X_1 = 2 - X_3 = 2 - C$$

Since C can be given any value, the original system of equations has an infinite number of solutions.

Exercises

The reader is encouraged to go through some of the following questions and exercises.

Chapter 1

1. Explain, in your own words, the meaning of "linear relationships."
2. What are the three components of a typical linear-programming problem?
3. What does the term *model* mean to you?
4. How would you classify the linear-programming model? Does this give you any idea of the limitations of linear programming?
5. Solve Exercise 6 below, and identify the solution stages in terms of the various steps of analytical decision making discussed in Chapter 1.

Chapter 2

6. A manufacturing firm sells two types of plastic containers for household consumption. The firm's market potential for these containers is unlimited.

   Each of the containers must be processed through two different machines. The relevant data on machine capacities, processing times, and profit contribution per unit are given in Table E 6. Using the graphical method, determine the optimal product mix under the following conditions:

   a. For the data given in Table E 6.
   b. The profit contribution of container A becomes 15 cents per unit.
   c. Containers A and B yield negative profit contributions (losses) of 10 cents and 5 cents, respectively.
   d. The capacity of machine A is increased to 2,500 minutes.

Table E 6

Machine	Processing time, minutes, per unit of		Available capacity per time period, minutes
	Container A	Container B	
1	4	2	2,000
2	3	5	3,000
Profit contribution per unit..	20¢	10¢	

7. Graph the following inequalities:

$$2X + 6Y \leq 1,000$$

$$X + 4Y \leq 400$$

Interpret these inequalities in terms of the concepts of "candidates" and "resources" discussed in Chapter 1.

8. The following algebraic statements represent a typical production problem which can be solved by linear programming.

Maximize $3X + 4Y$ subject to

$$5X + 8Y \leq 2,000$$

$$3X + 10Y \leq 1,000$$

and $X \geq 0$, $Y \geq 0$.

Solve this problem by the graphical method, and give some physical meaning to the algebraic statements.

9. Graph the following inequalities:

$$-X + 2Y \geq 100$$

$$-3X + 4Y \leq 300$$

$$4X + 6Y \leq 600$$

$$X \geq 0$$

$$Y \geq 0$$

a. Shade the area enclosed by the above inequalities.
b. Draw a straight line joining *any* two points in this shaded area (polygon). Does this line lie completely within the shaded area?

10. A toy manufacturer makes two types of model plastic boats. The management of this firm is submitted the facts in Table E 10 con-

Table E 10

Process	Product		Available time, minutes
	X	Y	
Molding..........................	10	5	80
Sanding and painting..............	6	6	66
Assembling...........	5	6	90
Profit contribution per unit........	$1.20	$1.00	

cerning the two products.   Determine an optimal program for the production of these products for a given period of time.

11. Design a linear-programming problem involving three competing "candidates" and two "resources."   Solve this problem graphically. Why can we not solve all linear-programming problems graphically?

Chapter 3

12. Solve Exercise 6 by the systematic trial-and-error method.

13. What action is suggested by the signs of the coefficients of the variables (candidates) in the modified objective function at different solution stages of Exercise 12?   How would you relate the coefficients of the variables in the final objective function to the "marginal" concept in economic theory?

14. Solve the linear-programming problem of Table E 14 by the systematic trial-and-error method.

Table E 14

Resource	Processing requirement for product		Capacity constraint
	A	B	
I	4	10	200
II	5	2	300
Profit contribution per unit........	$60	$100	

15.   Solve the following linear-programming problem by the systematic trial-and-error method:
Minimize $4X + 6Y$ subject to

$$2X + 3Y \geq 60$$

$$4X + Y \geq 40$$

and $X \geq 0$, $Y \geq 0$.
   Give some physical meaning to the algebraic statements of this problem.   Does the procedure for minimization differ from the one employed in solving Exercise 6?   If so, why?

Chapter 4

16.   How does a determinant differ from a matrix?

17.   Consider the 3×3 matrix

$$A = \begin{bmatrix} 2 & 0 & -1 \\ 1 & 2 & 4 \\ 0 & 3 & 2 \end{bmatrix}$$

Express A as a set of row vectors; as a set of column vectors. Find the determinant of A.

18.   Consider the following matrices:

$$A = \begin{bmatrix} 3 & 2 & 0 \\ 1 & 2 & 3 \\ 4 & 0 & 1 \end{bmatrix} \qquad B = \begin{bmatrix} 2 & 4 \\ 3 & 0 \\ 1 & 2 \end{bmatrix}$$

$$X = \begin{bmatrix} 1 & 4 \\ 1 & 2 \\ 4 & 0 \end{bmatrix} \qquad Y = \begin{bmatrix} 2 & 4 & 0 \\ 1 & -4 & 2 \end{bmatrix}$$

$$R = \begin{bmatrix} 1 & 4 & 0 \\ 1 & 2 & 3 \\ -6 & 1 & 2 \end{bmatrix} \qquad S = \begin{bmatrix} 1 & 0 & 0 \\ 0 & 1 & 0 \\ 0 & 0 & 1 \end{bmatrix}$$

Find AB, BA, XY, YX, RS, SR, ABY, S^2, and $(A - R)S$.

19.   Consider the following matrices:

$$A = \begin{bmatrix} 1 & 2 & 0 \\ 0 & 1 & 3 \\ 0 & 2 & 8 \end{bmatrix} \qquad B = \begin{bmatrix} 1 & -8 & 3 \\ 0 & 4 & -\frac{3}{2} \\ 0 & -1 & \frac{1}{2} \end{bmatrix}$$

Find AB and BA.   What is the inverse of A?

20.   Consider the following matrices:

$$A = \begin{bmatrix} 3 & 0 & 1 \\ 2 & 4 & 3 \end{bmatrix} \qquad B = \begin{bmatrix} 3 & 7 & 4 \\ 5 & 3 & 3 \\ 1 & 1 & 2 \end{bmatrix}$$

$$C = \begin{bmatrix} 4 & 3 \\ 1 & 0 \\ 6 & 2 \end{bmatrix} \qquad D = \begin{bmatrix} 3 & 4 & 2 \\ 6 & 1 & 0 \end{bmatrix}$$

$$V_1 = \begin{bmatrix} 2 \\ 4 \\ 5 \end{bmatrix} \qquad V_2 = \begin{bmatrix} 5 & 0 & 3 \end{bmatrix}$$

Find the following:
a. $A + B$, $A + D$, $C - D$, CA, CB, BV_1, BV_2, $V_1 B$, and $V_1 A$.
b. $k_1 A$ and $A k_1$, where $k_1 = 2$.
c. A^T and B^T.

21.   Explain the meanings of linear independence, vector space, and a basis for a vector space.

22.   Express

$$P_0 = \begin{bmatrix} 10 \\ 4 \\ 6 \end{bmatrix}$$

as a linear combination of P_1, P_2, and P_3, where

$$P_1 = \begin{bmatrix} 1 \\ 0 \\ 0 \end{bmatrix} \qquad P_2 = \begin{bmatrix} 4 \\ 1 \\ 2 \end{bmatrix} \qquad P_3 = \begin{bmatrix} 2 \\ -1 \\ 4 \end{bmatrix} \qquad P_4 = \begin{bmatrix} 2 \\ 4 \\ 10 \end{bmatrix}$$

What can you say about linear independence among P_1, P_2, P_3, and P_4?   Comment on your answer.

23.   Distinguish between a minor and a cofactor; between the cofactor matrix and the adjoint matrix.

24.   Let

$$A = \begin{bmatrix} 2 & 0 & -1 \\ 1 & 2 & 4 \\ 0 & 3 & 2 \end{bmatrix}$$

a. Find $|A|$.
b. Find A_{cofactor} and A_{adj}.
c. Find A^{-1}.

25. Find solutions to the following systems of equations:

a. $2X_1 + 6X_2 = 15$
$6X_1 + 3X_2 = 18$

b. $2X_1 + 6X_2 - X_3 = 15$
$6X_1 + 3X_2 - X_3 = 19$

c. $5X_1 + 4X_2 + X_3 = 100$
$2X_2 + 4X_3 = 80$
$2X_1 + 6X_2 = 90$

d. $5X_1 + 15X_2 = 100$
$2X_2 + 4X_3 = 80$
$2X_1 + 6X_2 = 90$

e. Write the coefficient matrix for (c), and find its determinant.
f. Write the coefficient matrix for (d), and find its determinant.

26. Explain the differences among the following types of solutions to linear-programming problems: (a) feasible solution, (b) basic feasible solution, and (c) degenerate basic feasible solution.

27. What is the relationship between the optimal solution of a linear-programming problem and each of the three types of solutions listed in Exercise 26?

Chapter 5

28. Solve Exercise 6 by the vector method, and compare the different solution stages with the corresponding solution stages of Exercise 12.

29. Upon completing the construction of his home, Mr. Green discovers that 100 square feet of plywood scrap and 80 square feet of white-pine scrap are in usable form for the construction of tables and bookcases. It takes 16 square feet of plywood and 8 square feet of white pine to make a table; 12 square feet of plywood and 16 square feet of white pine are required to construct a bookcase. By selling the finished products to a local unpainted-furniture store, Mr. Green can realize a profit of $5 on each table and $4 on each bookcase. How may he most profitably use the leftover wood? Employ the vector method to solve this problem.

Chapter 6

30. At the refinery of the Northeastern Petroleum Company three grades of gasoline are produced: high test, regular, and below regular. To produce each grade of gasoline, a fixed proportion of straight gasoline, octane, and additives is needed. A gallon of high test requires 20 percent straight gasoline, 50 percent octane, and 30 percent additives; a gallon of regular requires 50 percent straight gasoline, 30 percent octane, and 20 percent additives; a gallon of below regular calls for 70 percent straight gasoline, 20 percent octane, and 10 percent additives. It is known that Northeastern Petroleum receives a profit of 6, 5, and 4 cents on the high test, regular, and below regular, respectively.

If the supplies of straight gasoline, octane, and additives are restricted to 6,000,000, 2,000,000, and 1,000,000 gallons, respectively, how much of each grade should be produced, in a given time period, to maximize profits and make the best use of the resources? Does this product mix make full use of the resources?

31. The Martin Company, a diversified manufacturer, has recently started to produce razor blades as one of its products. At present the firm makes two types of blades, the "conventional" type that lasts for only a few shaves and the "stainless" type that enjoys a considerably longer life. The conventional type requires 8 units of carbon steel and 2 units of alloy steel per 100 blades, whereas the stainless type requires 4 units of carbon steel and 6 units of alloy steel per 100 blades. As a result of a recent strike, the company has been left with an inventory of 24,000 units of carbon steel and 10,000 units of alloy steel to be employed in the production of these two types of blades. How may the Martin Company best use the two resources in order to maximize the profit on razor blades? Assume that a profit of $1.0 can be obtained on each 100 conventional blades, and $1.5 on each 100 stainless blades.

32. An electronics firm is undecided as to the most profitable mix for its products. The products now manufactured are transistors, resistors, and electron tubes, with a profit (per 100 units) of $10, $6, and $4, respectively. To produce a shipment of transistors containing 100 units requires 1 hour of engineering service, 10 hours of direct labor, and 2 hours of administrative service. To produce 100 resistors requires 1 hour, 4 hours, and 2 hours of engineering, direct labor, and administrative time, respectively. To produce one shipment of electron tubes (100 units) requires 1 hour of engineering, 5 hours of direct labor, and 6 hours of administration. There are

100 hours of engineering services available, 600 hours of direct labor, and 300 hours of administration. What is the most profitable product mix?

33. A manufacturer is faced with a decision as to the extent of finishing to perform on his products prior to sales. He may sell his products as (a) raw casting, (b) semifinished product, or (c) finished product.

Each category requires the following amounts of work: (a) for raw casting: 5 hours of casting, 1 hour of machining, 1 hour of plating; (b) for semifinished: 5 hours of casting, 4 hours of machining, 1 hour of plating; (c) for finished: 5 hours of casting, 4 hours of machining, 2 hours of plating. The cost per hour of production in each of the departments is the same. The profit per unit of sales is as follows: raw casting, $1.50; semifinished, $2.50; finished, $3.00. The manufacturer can sell all the items which he can produce. The casting department has a daily production capacity of 130 hours, the machining department has a daily capacity of 88 hours, and the plating department has a 40-hour daily capacity.

How many products in each of the three categories should be manufactured in order to maximize daily profits from production?

34. The Universal Electric Company manufactures heavy-duty appliances for household use. It also features different models to be built in when homes are constructed or remodeled. One appliance available to contractors comes in three models: the standard model, No. 520; the deluxe all chrome, No. 570; and the semicomplete, No. 1521. The No. 520 retails for $35, the No. 570 including attachments for $50, and the No. 1521 for $30. All three models contain the same materials, costing $1 per unit, in varying quantities as shown below:

Model No.	Materials needed, units
570	20
520	10
1521	8

All three models are processed through the foundry and the assembly shop. The nonunion foundry has a wage rate of $2 per hour, and the unionized assembly shop has a wage rate of $2.50 per hour. Model 1521 requires 3 labor hours in the foundry and 2 labor hours in assembly. Model 570 requires 5 hours and 2 hours, and model 520 requires 2 hours and 4 hours, respectively.

Each product must also bear a portion of the overhead cost. Every unit of each model requires a fixed number of overhead hours equal to the sum of the direct-labor hours necessary to produce it. Overhead cost is then apportioned at the rate of $1 per overhead hour. Because of an aggressive sales campaign, management expects to sell everything that can be produced. However, owing to limited plant facilities and competing products, management can allocate only 3,000 hours of foundry time and 1,600 hours of assembly time for overall production of these models. In addition, there are only 2,400 units of material in stock and only 4,200 hours of budgeted overhead time.

What is the most profitable allocation of the available resources to these models?

35. The manager of the production department of the Lehigh Steel Company, Inc., has complete control of all functions within his department. He is mainly responsible for the scheduling of production.

All production is for storage only, with fabrication to follow after a slight delay. The production schedule is thus unaffected by the pattern of orders, and the production manager is free to schedule production so as to maximize profit.

For the current production period the company is concerned only with the production of nickel and chromium alloy steels. The production department receives 15 tons of nickel daily, along with a like amount of chromium. From these two alloying agents the company can produce alloy types 2010, 2020, and 1020. Alloy 2010 requires 200 pounds of nickel and 100 pounds of chromium per ton. Alloy 2020 requires 200 pounds of nickel and 200 pounds of chromium. Alloy 1020 requires 100 pounds of nickel and 200 pounds of chromium. The alloys net $25, $30, and $35 profit per ton produced, respectively.

Assuming that 2 tons of alloy 2010 must be produced, design the most profitable production program.

Chapter 7

36. Minimize $4X + 6Y + Z$ subject to

$$X + 2Y \qquad \geq 10$$
$$Y + 4Z \geq 20$$
$$3X \qquad + Z \geq 40$$

and $X \geq 0$, $Y \geq 0$, $Z \geq 0$.   Give some physical interpretation to the above algebraic statements.

37. West Chester, a small eastern city of 15,000 people, requires an average of 300,000 gallons of water daily.   The city is supplied from a central waterworks, where the water is purified by such conventional methods as filtration and chlorination.   In addition, two different chemical compounds, softening chemical and health chemical, are needed for softening the water and for health purposes. The waterworks plans to purchase two popular brands that contain these chemicals.   One unit of the Chemco Corporation's product gives 8 pounds of the softening chemical and 3 pounds of the health chemical.   One unit of the American Chemical Company's product contains 4 pounds and 9 pounds per unit, respectively, for the same purposes.

To maintain the water at a minimum level of softness and to meet a minimum in health protection, experts have decided that 150 and 100 pounds of the two chemicals that make up each product must be added to the water daily.   At a cost of $8 and $10 per unit, respectively, for Chemco's and American Chemical's products, what is the optimal quantity of each product that should be used to meet the minimum level of softness and a minimum health standard?

38. The H.E.E. Construction Company is building roads on the side of South Mountain.   It is necessary to use explosives to blow up the boulders under the ground to make the surface level.   There are three liquid ingredients (A, B, C) in the liquid explosive used.   It is known that at least 10 ounces of the explosive has to be used to get results.   If more than 20 ounces is used, the explosion will be too damaging.   Also, to have an explosion, at least $\frac{1}{4}$ ounce of ingredient C must be used for every ounce of ingredient A, and at least 1 ounce of ingredient B must be used for every ounce of ingredient C.   The costs of ingredients A, B, and C are $6, $18, and $20 per ounce, respectively.   Find the least-cost explosive mix necessary to produce a safe explosion.

39. Minimize $f(X) = 30X_1 + 20X_2$ subject to

$$X_1 + X_2 \leq 8$$
$$6X_1 + 4X_2 \geq 12$$
$$5X_1 + 8X_2 = 20$$

and $X_1 \geq 0$, $X_2 \geq 0$.

40.   Minimize $f(X) = X_1 - X_2 + 2X_3$ subject to

$$4X_1 - 2X_2 + 3X_3 \geq 12$$
$$-2X_2 + X_3 \geq 8$$
$$3X_1 + X_2 + 6X_3 \geq 18$$

and $X_1 \geq 0$, $X_2 \geq 0$, $X_3 \geq 0$.

Chapter 8

41.   Form the dual to Exercise 6.   What economic significance can be attached to the dual variables?

42.   A small winery manufactures two types of wine, Burbo's Better and Burbo's Best.   Burbo's Better sells for $20 per quart, whereas Burbo's Best sells for $30 per quart.   Two men mix the two wines. It takes a man 2 hours to mix a quart of Burbo's Better and 6 hours to mix a quart of Burbo's Best.   Each man works an 8-hour day, with 30 minutes for lunch.   The quantity of alcohol that the plant can use is limited to 18 ounces daily.   Six ounces is used in each quart of Burbo's Better, and 3 ounces in each quart of Burbo's Best.   How many quarts of each should the winery make?   Solve this problem via its dual.

43.   Wung Lee's Chinese Restaurant serves two types of chow mein, chicken and shrimp.   There is a $2.00 profit on each pound of shrimp chow mein, and a $1.00 profit on chicken chow mein.   From past experience Wung Lee knows that he should not make more than 10 pounds of chicken chow mein and 8 pounds of shrimp chow mein each day.   The two cooks work an 8-hour day, with 30 minutes for lunch.   A pound of chicken chow mein takes 2 hours to cook, whereas a pound of shrimp chow mein takes 1 hour.   How many pounds of each type of chow mein should Wung Lee make?   Utilize the dual approach in solving this problem.

44.   The Catchem Every Time Company manufactures two types of baseball gloves.   The profit from the high-quality glove is $4.00, and the profit from the low-quality glove is $2.00.   Machine A is the only machine on which these gloves can be made.   It takes 30 minutes to manufacture a low-quality glove, whereas 2 hours is needed to make a high-quality glove.   The plant works two 8-hour shifts.   There is enough material to make 24 low-quality gloves per day.   It takes twice as much material to manufacture a high-quality glove. How many of each type of glove should the company make? Form the dual, and obtain the values of the dual variables from a regular simplex solution of this problem.

Chapter 9

45. How can we detect degeneracy during the solution stages of a linear-programming problem?
46. What can be said for the solution vector of a degenerate problem having n structural variables and m structural constraints? What happens if a linear-programming problem begins to cycle?
47. Maximize $f(X) = 10X_1 + 30X_2 + 25X_3$ subject to

$$5X_1 + 4X_2 + 10X_3 \leq 40$$
$$3X_1 + 2X_2 + 9X_3 \leq 25$$
$$X_1 - X_2 - 3X_3 \leq 10$$

and $X_1 \geq 0$, $X_2 \geq 0$, $X_3 \geq 0$.

48. The Roadhog Truck Company manufactures two truck types, the Snort and the Razorback. The Snort is a 50-ton truck, whereas the Razorback is 40 tons. The company has unlimited demand for both trucks, but plant facilities limit production. Each truck must pass through the three departments of the plant. The man-hours needed for each truck and the total man-hours available per month are given in Table E 48. The profit from each Snort is $2,000, and

Table E 48

Department	Man-hours needed per truck		Man-hours available per month
	Snort	Razorback	
Motor..........	30	40	1,000
Assembly........	20	11	275
Painting........	4	5	335

the profit from each Razorback is $2,600. How should monthly production be scheduled to maximize profits?

49. The Wild Horses Oil Company makes three brands of gasoline: Man o' War, Trigger, and Swayback. Wild Horses makes its products by blending two grades of gasoline, each with a different octane rating. Each brand of gas must have an octane rating greater than a predetermined minimum. In gasoline blending, final octane rating is linearly proportional to component octanes (i.e., a blend of 50 percent 100-octane and 50 percent 200-octane gasoline is 150 octane). The other relevant data are given in Table E 49a and b. There is no limit on the amount of each gasoline

Table E 49a

Blending component	Octane	Cost per gallon, cents	Supply per week, gallons
A	200	10	20,000
B	130	8	10,000

Table E 49b

Brand	Minimum octane	Sale price per gallon, cents
Man o' War.........	180	24
Trigger..............	160	21
Swayback...........	140	12

that may be sold. Determine weekly production to maximize profits.

Chapter 10

50. Keller and Hogan Enterprises have manufacturing plants located in Newark, Atlanta, Cleveland, and Denver. Their market demands necessitate warehouses in Bethlehem, Chicago, Dallas, Kansas City, and Seattle. Let the matrix in Table E 50 denote the maximum capacities of the respective plants and the demand requirements of the various warehouses. The small number in the right-hand corner of each cell represents the cost of shipping 1 unit from the particular plant to the given warehouse. Find the least-cost shipping assignment.

Table E 50

Plant	Warehouse					Plant capacity
	B	C	D	K	S	
Newark	1	2	6	2	3	800
Atlanta	3	4	5	8	1	600
Cleveland	3	1	1	2	6	200
Denver	4	7	3	5	4	400
Warehouse demand	400	100	700	300	500	

51. Solve Exercise 50 after making the initial assignment by Vogel's approximation method.

52. Explain, in your own words, the parallel points in the transportation model and the general linear-programming model.

53. Given the transportation-cost matrix in Table E 53, minimize the total cost of transportation consistent with the capacities of the origins and the requirements of the destinations.

54. Assume that the payoffs in Exercise 53 represent profits. What, then, is the optimal shipping program?

55. What happens if the initial assignment in a transportation problem gives less than $m + n - 1$ occupied cells, where m represents the number of origins, and n represents the number of destinations? How do we handle such a situation?

Chapter 11

56. The matrix in Table E 56 gives the cost of assigning a particular job to a specific machine. Make the necessary assignments to accomplish all the jobs at a minimum total cost.

57. Assume that the numbers in Table E 56 represent profits. Will the optimal assignment change?

Table E 53

Origin	Destination			Capacity per time period
	D_1	D_2	D_3	
O_1	3	2	4	100
O_2	0	4	2	300
O_3	2	1	3	400
O_4	5	2	3	200
O_5	1	4	0	300
O_6	4	1	2	300
O_7	1	6	4	400
Requirement per time period	700	900	400	

Table E 56

Job	Machine		
	A	B	C
1	$380	$610	$330
2	210	380	415
3	260	210	300

58. The Lehigh Company is faced with the problem of assigning production orders for the next month to specific departments. Knowing how much it costs each department to produce each order, the company must determine the lowest cost of overall production when only one job is assigned to each department. Starting with the matrix in Table E 58, explain how you would arrive at your final solution. Obtain the solution.

Table E 58

Production order	Department				
	A	B	C	D	E
1	$120	$150	$ 75	$ 90	$100
2	140	80	90	85	170
3	50	40	40	70	110
4	75	65	45	70	90
5	110	90	140	115	100

59. Solve Exercise 56 by employing the mathematical approach of Section 11.11. What do the u_i's and v_j's represent?

60. Why do we face an inherently degenerate situation in the assignment model?

Selected
Bibliography

Albers, Henry H.: "Organized Executive Action: Decision-making, Communication, and Leadership," John Wiley & Sons, Inc., New York, 1961.

Bowman, E. H.: Production Scheduling by the Transportation Method of Linear Programming, *Operations Research*, vol. 4, no. 1, February, 1956.

Charnes, A., W. W. Cooper, and A. Henderson: "An Introduction to Linear Programming," John Wiley & Sons, Inc., New York, 1953.

Chung, An-Min: "Linear Programming," Charles E. Merrill Books, Inc., Columbus, Ohio, 1963.

Churchman, C. West, R. L. Ackoff, and E. L. Arnoff: "Introduction to Operations Research," John Wiley & Sons, Inc., New York, 1957.

Dean, Burton V.: Applications of Operations Research to Managerial Decision-making, *Administrative Science Quarterly*, vol. 3, no. 3, December, 1958.

Dorfman, Robert: Operations Research, *American Economic Review*, vol. 50, no. 4, September, 1960.

————, Paul A. Samuelson, and Robert M. Solow: "Linear Programming and Economic Analysis," McGraw-Hill Book Company, New York, 1958.

Ferguson, Robert O., and L. F. Sargent: "Linear Programming," McGraw-Hill Book Company, New York, 1958.

Forrester, Jay W.: "Industrial Dynamics," The M.I.T. Press, Cambridge, Mass., and John Wiley & Sons, Inc., New York, 1961.

Fuller, Leonard E.: "Basic Matrix Theory," Prentice-Hall, Inc., Englewood Cliffs, N.J., 1962.

Garvin, W. W.: "Introduction to Linear Programming," McGraw-Hill Book Company, New York, 1960.

Graves, Robert L., and Philip Wolfe: "Recent Advances in Mathematical Programming," McGraw-Hill Book Company, New York, 1963.

279

280 Linear Programming

Manne, Alan S.: "Economic Analysis for Business Decisions," McGraw-Hill Book Company, New York, 1961.

Miller, David W., and Martin K. Starr: "Executive Decisions and Operations Research," Prentice-Hall, Inc., Englewood Cliffs, N.J., 1960.

Morse, Phillip M., and G. E. Kimball: "Methods of Operations Research," John Wiley & Sons, Inc., New York, 1951.

Raymond, Fairfield E.: "Quantity and Economy in Manufacture," McGraw-Hill Book Company, New York, 1931.

Reinfeld, N. V., and William R. Vogel: "Mathematical Programming," Prentice-Hall, Inc., Englewood Cliffs, N.J., 1958.

Saaty, T. L.: "Mathematical Methods of Operations Research," McGraw-Hill Book Company, New York, 1959.

Sasieni, Maurice, A. Yaspan, and L. Friedman: "Operations Research: Methods and Problems," John Wiley & Sons, Inc., New York, 1960.

Shuchman, Abe: "Scientific Decision-making in Business," Holt, Rinehart and Winston, Inc., New York, 1963.

Vance, Stanley: "Industrial Administration," McGraw-Hill Book Company, New York, 1959.

Von Neumann, John, and O. Morgenstern: "Theory of Games and Economic Behavior," Princeton University Press, Princeton, N.J., 1947.

Index

281